AF531601

Ocean Engineering

Ocean Engineering

Ravinder Bora

RANDOM PUBLICATIONS
NEW DELHI (INDIA)

Ocean Engineering

ISBN 978-93-5111-404-8

Published in 2014 in India by

RANDOM PUBLICATIONS

4376-A/4B, Gali Murari Lal, Ansari Road
New Delhi-110 002
Phone : +91-11-43580356, +91-11-23289044
e-mail: randomexports@gmail.com, sales@randompublications.com,
info@randompublications.com

Type Setting by : Keystoneprintads, Delhi-110051
Printed at Thomson Press (India) Ltd

Preface

The field of ocean engineering provides an important link between the other oceanographic disciplines such as marine biology, chemical and physical oceanography, and marine geology and geophysics. Just as the interests of oceanographers have driven the demand for the design skills and technical expertise of ocean engineers, the innovations in instrumentation and equipment design made by ocean engineers have revolutionized the field of oceanography.

Ocean Engineering is a multidisciplinary engineering field aimed at solving engineering problems associated with working in the ocean environment and wisely exploring and harnessing the ocean's resources. Ocean engineers design, build, operate and maintain ships, offshore structures and ocean technologies as diverse as aircraft carriers, submarines, sailboats, tankers, tugboats, yachts, oil rigs, underwater robots, and acoustic sonar. Ocean engineers study the oceans to determine the effects of waves, currents, and the salt water environment on ships and other marine vehicles and structures.

They develop methods and materials to withstand wave forces and protect against corrosion. Ocean engineers are involved in the development and use of manned and remotely operated sub-surface vehicles for deep-sea exploration and resource recovery. Ocean engineering majors learn how to use science and math to develop ports, platforms, dikes, and other ocean structures, with the exception of ships. Areas of study include acoustics (the science of sound), measurement, the behavior of waves and sediments, and much more.

Ocean Engineering is theprogram that prepares individuals to apply mathematical and scientific principles to the design, development and operational evaluation of systems to monitor, control, manipulate and operate within coastal or ocean environments, such as underwater platforms, flood control systems, dikes, hydroelectric power systems, tide and current control and warning systems, and communications equipment; the planning and design of total systems for working and functioning in water or underwater environments; and the analysis of related engineering problems such as the action of water properties and behavior on physical systems and people, tidal forces, current movements, and wave motion.

It is envisaged that the book would serve as a reference material to the future researchers and also provide basic information to all the concerned viz. the community, the students, the industry, media. To benefit a wide range of readers the book has been written in simple and plain English, with appropriate acronyms and glossary and the key facts or summary given at the beginning of each chapter.

I thank all members of my team who have helped in the preparation of the book. My special thanks go to "Random Publications" who have published the book.

*– **Ravinder Bora***

Contents

1

Ocean and Engineering

An ocean; the World Ocean of classical antiquity) is a body of saline water that composes a large part of a planet's hydrosphere. In the context of Earth, it refers to one or all of the major divisions of the planet's World Ocean – they are, in descending order of area, the Pacific, Atlantic, Indian, Southern (Antarctic), and Arctic Oceans. The word *sea* is often used interchangeably with "ocean" in American English but, strictly speaking, a sea is a body of saline water (generally a division of the World Ocean) partly or fully enclosed by land. The general characteristics of seas and oceans are covered at sea. Earth is the only known planet to have an ocean (or any large amounts of open liquid water).

Approximately 72% of the planet's surface (~3.6×10^8 km^2) is covered by saline water that is customarily divided into several principal oceans and smaller seas, with the ocean covering approximately 71% of the Earth's surface. The ocean contains 97% of the Earth's water, and oceanographers have stated that only 5% of the ocean as a whole on Earth has been explored. The total volume is approximately 1.3 billion cubic kilometres (310 million cu mi) with an average depth of 3,682 metres (12,080 ft).

Because it is the principal component of Earth's hydrosphere, the world ocean is integral to all known life, forms part of the carbon cycle, and influences climate and weather patterns. It is the habitat of 230,000 known species, although much of the ocean's depths remain unexplored and it is estimated that over two million marine species exist. The origin of Earth's oceans is still unknown, but oceans are believed to have formed in the Hadean period and may have been the impetus for the emergence of life.

Extraterrestrial oceans may be composed of water or other elements and compounds. The only confirmed large stable bodies of extraterrestrial surface liquids are the lakes of Titan, although there is evidence for the existence of oceans elsewhere in the Solar System. Early in their geologic histories, Mars and Venus are theorized to have had large water oceans. The Mars ocean hypothesis suggests that nearly a third of the surface of Mars was once covered by water, and a runaway greenhouse effect may have boiled away the global ocean of Venus. Compounds such as salts and ammonia dissolved in water lower its freezing point, so that water might exist in large quantities in extraterrestrial environments as brine or convecting ice. Unconfirmed oceans are speculated beneath the surface of many dwarf planets and natural satellites; notably, the ocean of Europa is believed to have over twice the water volume of Earth. The Solar System's gas giant planets are also believed to possess liquid atmospheric layers of yet to be confirmed compositions. Oceans may also exist on exoplanets and exomoons, including surface oceans of liquid water within a circumstellar habitable zone. Ocean planets are a hypothetical type of planet with a surface completely covered with liquid.

EARTH'S GLOBAL OCEAN

DIVISIONS

Though generally described as several separate oceans, these waters comprise one global, interconnected body of salt water sometimes referred to as the World Ocean or global ocean. This concept of a continuous body of water with relatively free interchange among its parts is of fundamental importance to oceanography. The major oceanic divisions are defined in part by the continents, various archipelagos, and other criteria. See the table below for more information; note that the table is in descending order in terms of size. The Pacific and Atlantic may be further subdivided by the equator into northern and southern portions. A smaller region of the ocean can be called other names, such as sea, gulf, bay, and strait.

PHYSICAL PROPERTIES

The total mass of the hydrosphere is about 1,400,000,000,000,000,000 metric tons (1.5×1018 short tons) or 1.4×1021 kg, which is about 0.023 percent of the Earth's total mass. Less than 3 percent is freshwater; the rest is saltwater, mostly in the ocean. The area of the World Ocean is 361 million square kilometres (139 million square miles), and its volume is approximately 1.3 billion cubic kilometres (310 million cu mi). This can be thought of as a cube of water with an edge length of 1,111 kilometres (690 mi). Its average depth is 3,790 metres (12,430 ft), and its maximum depth is 10,923 metres (6.787 mi). Nearly half of the world's marine waters are over 3,000 metres (9,800 ft) deep. The vast expanses of deep ocean (anything below 200 metres (660 ft)) cover about 66% of the Earth's surface. This does not include seas not connected to the World Ocean, such as the Caspian Sea.

The bluish color of water is a composite of several contributing agents. Prominent contributors include dissolved organic matter and chlorophyll. Sailors and other mariners have reported that the ocean often emits a visible glow, or luminescence, which extends for miles at night. In 2005, scientists announced that for the first time, they had obtained photographic evidence of this glow. It is most likely caused by bioluminescence.

ZONES AND DEPTHS

Oceanographers divide the ocean into different zones depending on the present physical and biological conditions. The pelagic zone includes all open ocean regions, and can be divided into further regions categorized by depth and light abundance. The photic zone covers the oceans from surface level to 200 metres down. This is the region where photosynthesis can occur and therefore is the most biodiverse.

Since plants require photosynthesis, life found deeper than this must either rely on material sinking from above (see marine snow) or find another energy source; hydrothermal vents are the primary option in what is known as the aphotic zone (depths exceeding 200 m). The pelagic part of the photic zone is known as the epipelagic.

The pelagic part of the aphotic zone can be further divided into regions that succeed each other vertically according to temperature. The mesopelagic is the uppermost region. Its lowermost boundary is at a thermocline of 12 °C (54 °F), which, in the tropics generally lies at 700-1,000 metres (2,300-3,300 ft). Next is the bathypelagic lying between 10 and 4 °C (50 and 39 °F), typically between 700-1,000 metres (2,300-3,300 ft) and 2,000-4,000 metres (6,600-13,000 ft) Lying along the top of the abyssal plain is the abyssopelagic, whose lower boundary lies at about 6,000 metres (20,000 ft). The last zone includes the deep oceanic trench, and is known as the hadalpelagic. This lies between 6,000-11,000 metres (20,000-36,000 ft) and is the deepest oceanic zone.

Along with pelagic aphotic zones there are also benthic aphotic zones. These correspond to the three deepest zones of the deep-sea. The bathyal zone covers the continental slope down to about 4,000 metres (13,000 ft). The abyssal zone covers the abyssal plains between 4,000 and 6,000 m. Lastly, the hadal zone corresponds to the hadalpelagic zone which is found in the oceanic trenches. The pelagic zone can also be split into two subregions, the neritic zone and the oceanic zone. The neritic encompasses the water mass directly above the continental shelves, while the oceanic zone includes all the completely open water. In contrast, the littoral zone covers the region between low and high tide and represents the transitional area between marine and terrestrial conditions. It is also known as the intertidal zone because it is the area where tide level affects the conditions of the region.

EXPLORATION

Ocean travel by boat dates back to prehistoric times, but only in modern times has extensive underwater travel become possible. The deepest point in the ocean is the Mariana Trench, located in the Pacific Ocean near the Northern Mariana Islands. Its maximum depth has been estimated to be 10,971 metres (35,994 ft) (plus or minus 11 meters; see the Mariana Trench article for discussion of the various estimates of the maximum depth.) The British naval vessel, Challenger II surveyed the trench in 1951 and named the deepest part of the trench, the "Challenger Deep". In 1960, the Trieste successfully reached the bottom of the trench, manned by a crew of two men. Much of the ocean bottom remains unexplored and unmapped. A global image of many underwater features larger than 10 kilometres (6.2 mi) was created in 1995 based on gravitational distortions of the nearby sea surface.

CLIMATE

Ocean currents greatly affect the Earth's climate by transferring heat from the tropics to the polar regions. transferring warm or cold air and precipitation to coastal regions, where winds may carry them inland. Surface heat and freshwater fluxes create global density gradients that drive the thermohaline circulation part of large-scale ocean circulation. It plays an important role in supplying heat to the polar regions, and thus in sea ice regulation. Changes in the thermohaline circulation are thought to have significant impacts on the Earth's radiation budget. Insofar as the thermohaline circulation governs the rate at which deep waters reach the surface, it may also significantly influence atmospheric carbon dioxide concentrations.

It is often stated that the thermohaline circulation is the primary reason that the climate of Western Europe is so temperate. An alternate hypothesis claims that this is largely incorrect, and that Europe is warm mostly because it lies downwind of an ocean basin, and because atmospheric waves bring warm air north from the subtropics. The Antarctic Circumpolar Current encircles that continent, influencing the area's climate and connecting currents in several oceans. One of the most dramatic forms of weather occurs over the oceans: tropical cyclones (also called "typhoons" and "hurricanes" depending upon where the system forms).

BIOLOGY

The ocean has a significant effect on the biosphere. Oceanic evaporation, as a phase of the water cycle, is the source of most rainfall, and ocean temperatures determine climate and wind patterns that affect life on land. Life within the ocean evolved 3 billion years prior to life on land. Both the depth and the distance from shore strongly influence the biodiversity of the plants and animals present in each region.

Lifeforms native to the ocean include:

- Fish;
- Radiata, such as jellyfish (Cnidaria);
- Cetacea, such as whales, dolphins, and porpoises;
- Cephalopods, such as octopus and squid;
- Crustaceans, such as lobsters, clams, shrimp, and krill;
- Marine worms;
- Plankton; and
- Echinoderms, such as brittle stars, starfish, sea cucumbers, and sand dollars.

ECONOMIC VALUE

The oceans are essential to transportation. This is because most of the world's goods move by ship between the world's seaports. Oceans are also the major supply source for the fishing industry. Some of the more major ones are shrimp, fish, crabs and lobster.

EXTRATERRESTRIAL OCEANS

While Earth is the only known planet with large stable bodies of liquid water on its surface and the only one in our Solar System, other celestial bodies are believed to possess large oceans.

PLANETS

The gas giants, Jupiter and Saturn, are thought to lack surfaces and instead have a stratum of liquid hydrogen, however their planetary geology is not well understood. The possibility of Uranus and Neptune possessing hot, highly compressed, supercritical water under their thick atmospheres has been hypothesised. While their composition is still not fully understood, a 2006 study by Wiktorowicz et al. ruled out the possibility of such an water "ocean" existing on Neptune, though some studies have suggested that exotic oceans of liquid diamond are possible. The Mars ocean hypothesis suggests that nearly a third of the surface of Mars was once covered by water, though the water on Mars is no longer oceanic. The possibility continues to be studied along with reasons for their apparent disappearance. Astronomers believe that Venus had liquid water and perhaps oceans in its very early history. If they existed, all later vanished via resurfacing.

NATURAL SATELLITES

A global layer of liquid water thick enough to decouple the crust from the mantle is believed to be present on Titan, Europa and, with less certainty, Callisto, Ganymede and Triton. A magma ocean is thought to be present on Io. Geysers have been found on Saturn's moon Enceladus, though their origins are not well understood. Other icy moons may also have internal oceans, or have once had internal oceans that have now frozen.

Large bodies of liquid hydrocarbons are thought to be present on the surface of Titan, though they are not large enough to be described as oceans and are sometimes referred to as lakes or seas. The Cassini-Huygens space mission initially discovered only what appeared to be dry lakebeds and empty river channels, suggesting that Titan had lost what surface liquids it might have had. Cassini's more recent fly-by of Titan offers radar images that strongly suggest hydrocarbon lakes exist near the colder polar regions. Titan is thought to have a subterranean water ocean under the ice and hydrocarbon mix that forms its outer crust.

DWARF PLANETS AND TRANS-NEPTUNIAN OBJECTS

Ceres appears to be differentiated into a rocky core and icy mantle and may harbour a liquid-water ocean under its surface. Not enough is known of the larger Trans-Neptunian objects to determine whether they are differentiated bodies capable of possessing oceans, although models of radioactive decay suggest that Pluto, Eris, Sedna, and Orcus have oceans beneath solid icy crusts at the core-boundary approximately 100 to 180 km thick.

EXTRASOLAR

Some planets and natural satellites beyond the Solar System are likely to possess oceans, including possible water ocean planets similar to Earth in the habitable zone or "liquid-water belt". The detection of oceans, even through the spectroscopy method, however is likely to prove extremely difficult and inconclusive. Theoretical models have been used to predict with high probability that GJ 1214 b, detected by transit, is composed of exotic form of ice VII, making up 75% of its mass.

Other possible candidates are merely speculated based on their mass and position in the habitable zone include planet though little is actually known of their composition. Some scientists speculate Kepler-22b may be an "ocean-like" planet. Models have been proposed for Gliese 581 d that could include surface oceans. Gliese 436 b is speculated to have an ocean of "hot ice". Extrasolar moons orbiting planets, particularly gas giants within their parent star's habitable zone may theoretically possess surface oceans.

OCEAN ENGINEERING

Ocean or Marine engineering broadly refers to the engineering of boats, ships, oil rigs and any other marine vessel or structure. Specifically, marine engineering is the discipline of applying engineering sciences, mostly mechanical and electrical engineering, to the development, design, operation and maintenance of watercraft propulsion and on-board systems; e.g. power and propulsion plants, machinery, piping, automation and control systems etc. for marine vehicles of any kind like surface ships, submarines etc.

- The engineering of a vessel's propulsion system.
- The engineering of shipboard systems and machinery.
- A ship's engineering department, an organizational unit that is responsible for running the vessel's propulsion systems and support systems for crew, passengers and cargo; this field career track within marine engineering is, more specifically, referred to as seagoing engineering.
- In limited and specific ship-related context, the engineering of structures to support vessels.
- Oceanographic engineering, also called marine electronics

engineering, is concerned with the design of electronic devices for use in the marine environment, such as the remote sensing systems used by oceanographers.

Not all marine engineering is concerned with moving vessels. Offshore construction, also called offshore engineering, ocean engineering or maritime engineering, is concerned with the technical design of fixed and floating marine structures, such as oil platforms and offshore wind farms.

2

Harbor and Ports

HARBOR

A harbor or harbour (see spelling differences), or haven, is a body of water where ships, boats, and barges can seek shelter from stormy weather, or else are stored for future use. Harbors can be natural or artificial. An artificial harbor has deliberately constructed breakwaters, sea walls, or jettys, or otherwise, they could have been constructed by dredging, and these require maintenance by further periodic dredging. An example of the former kind is at Long Beach Harbor, California, and an example of the latter kind is San Diego Harbor, California, which was, under natural conditions, too shallow for modern merchant ships and warships.

In contrast, a natural harbor is surrounded on several sides by prominences of land. An example of this kind of harbor is San Francisco Bay, California. Harbors and ports are often confused with each other. A port is a facility for loading and unloading vessels; ports are usually located in harbors.

ARTIFICIAL HARBORS

Artificial harbors are frequently built for use as ports. The oldest artificial harbor known is the Ancient Egyptian site at Wadi al-Jarf, on the Red Sea coast, which is at least 4500 years old (ca. 2600-2550 BC, reign of King Khufu). The largest artificially created harbor is Jebel Ali in Dubai. Other large and busy artificial harbors are located at: Rotterdam, The Netherlands; Port of Houston, Texas; Port of Long Beach, California; and Port of Los Angeles in San Pedro, California. The Ancient Carthaginians constructed fortified, artificial harbors called cothons.

NATURAL HARBORS

A natural harbor is a landform where a part of a body of water is protected and deep enough to furnish anchorage. Many such harbors are rias. Natural harbors have long been of great strategic naval and economic importance, and many great cities of the world are located on them. Having a protected harbor reduces or eliminates the need for breakwaters as it will result in calmer waves inside the harbor. Some examples are New York Harbor in the United States; Poole Harbour in England; Kingston Harbour in Jamaica; Subic, Grand Harbour in Malta, Zambales in the Philippines; Sydney Harbour in Australia; Pearl Harbor in Hawaii; San Francisco Bay in California; Visakhapatnam Harbour in Andhra Pradesh, India; Killybegs in County Donegal Ireland; and Halifax Harbour in Nova Scotia, Canada.

ICE-FREE HARBORS

For harbors near the North and South Poles, being ice-free is an important advantage, especially when it is year-round. Examples of these include Murmansk, Russia; Pechenga, Russia; St. Petersburg, Russia; Hammerfest, Norway; Vardø, Norway; and Prince Rupert Harbour, Canada. The world's southmost harbor, located at Antarctica's Winter Quarters Bay (77° 50? South), is potentially ice-free, depending on the summertime pack ice conditions.

PORT

A port is a location on a coast or shore containing one or more harbors where ships can dock and transfer people or cargo to or from land. Port locations are selected to optimize access to land and navigable water, for commercial demand, and for shelter from wind and waves. Ports with deeper water are rarer, but can handle larger, more economical ships. Since ports throughout history handled every kind of traffic, support and storage facilities vary widely, may extend for miles, and dominate the local economy. Some ports have an important military role.

DISTRIBUTION

Ports often have cargo-handling equipment, such as cranes (operated by longshoremen) and forklifts for use in loading ships, which may be provided by private interests or public bodies. Often, canneries or other processing facilities will be located nearby. Some ports feature canals, which allow ships further movement inland. Access to intermodal transportation, such as trains and trucks, are critical to a port, so that passengers and cargo can also move further inland beyond the port area. Ports with international traffic have customs facilities. Harbour pilots and tugboats may maneuver large ships in tight quarters when near docks.

TYPES

The terms "port" and "seaport" are used for different types of port facilities that handle ocean-going vessels, and river port is used for river traffic, such as barges and other shallow-draft vessels.

Inland port

An Inland port is a port on a navigable lake, river (fluvial port), or canal with access to a sea or ocean, which therefore allows a ship to sail from the ocean inland to the port to load or unload its cargo.

Fishing port

A fishing port is a port or harbour for landing and distributing fish. It may be a recreational facility, but it is usually commercial. A fishing port is the only port that depends on an ocean product, and depletion of fish may cause a fishing port to be uneconomical. In recent decades, regulations to save fishing stock may limit the use of a fishing port, perhaps effectively closing it.

Dry port

A dry port is an inland intermodal terminal directly connected by road or rail to a seaport and operating as a centre for the transshipment of sea cargo to inland destinations.

Warm-water port

A warm water port is one where the water does not freeze in winter time. Because they are available year-round, warm water ports can be of great geopolitical or economic interest. Such settlements as Vostochny Port, Murmansk and Petropavlovsk-Kamchatsky in Russia, Odessa in Ukraine, Kushiro in Japan and Valdez at the terminus of the Alaska Pipeline owe their very existence to being ice-free ports.

Seaport

A seaport is further categorized as a "cruise port" or a "cargo port". Additionally, "cruise ports" are also known as a "home port" or a "port of call". The "cargo port" is also further categorized into a "bulk" or "break bulk port" or as a "container port".

Cruise home port

A cruise home port is the port where cruise-ship passengers board (or embark) to start their cruise and disembark the cruise ship at the end of their cruise. It is also where the cruise ship's supplies are loaded for the cruise, which includes everything from fresh water and fuel to fruits, vegetable, champagne, and any other supplies needed for the cruise. "Cruise home ports" are a very busy place during the day the cruise ship is in port, because off-going passengers debark their baggage and on-coming passengers board the ship in addition to all the supplies being loaded. Currently, the Cruise Capital of the World is the Port of Miami, Florida, closely followed behind by Port Everglades, Florida and the Port of San Juan, Puerto Rico.

Port of call

A port of call is an intermediate stop for a ship on its sailing itinerary, which may include up to half a dozen ports. At these ports, a cargo ship may take on supplies or fuel, as well as unloading and loading cargo. But for a cruise ship, it is their premier stop where the cruise lines take on passengers to enjoy their vacation.

Cargo port

Cargo ports, on the other hand, are quite different from cruise ports, because each handles very different cargo, which has to be loaded and unloaded by very different mechanical means. The port may handle one particular type of cargo or it may handle numerous cargoes, such as grains, liquid fuels, liquid chemicals, wood, automobiles, etc. Such ports are known as the "bulk" or "break bulk ports". Those ports that handle containerized cargo are known as container ports. Most cargo ports handle all sorts of cargo, but some ports are very specific as to what cargo they handle. Additionally, the individual cargo ports are divided into different operating terminals which handle the different cargoes, and are operated by different companies, also known as terminal operators or stevedores.

ACCESS

Ports sometimes fall out of use. Rye, East Sussex, was an important English port in the Middle Ages, but the coastline changed and it is now 2 miles (3.2 km) from the sea, while the ports of Ravenspurn and Dunwich have been lost to coastal erosion. Also in the United Kingdom, London, on the River Thames, was once an important international port, but changes in shipping methods, such as the use of containers and larger ships, put it at a disadvantage.

ENVIRONMENTAL EFFECT

There are several initiatives to decrease ports' environmental effect. These include the World Ports Climate Initiative, the African Green Port Initiative and EcoPorts.

MARINE PROPULSION

Marine propulsion is the mechanism or system used to generate thrust to move a ship or boat across water. While paddles and sails are still used on some smaller boats, most modern ships are propelled by mechanical systems consisting of a motor or engine turning a propeller, or less frequently, in jet drives, an impeller. Marine engineering is the discipline concerned with the design of marine propulsion systems. Steam engines were the first mechanical engines used in marine propulsion, but have mostly been replaced by two-stroke or four-stroke diesel engines, outboard motors, and gas turbine engines on faster ships. Nuclear reactors producing steam are used to propel warships and icebreakers, and there have been attempts to utilize them to power commercial vessels. Electric motors have been used on submarines and electric boats and have been proposed for energy-efficient propulsion. Recent development in liquified natural gas (LNG) fueled engines are gaining recognition for their low emissions and cost advantages.

POWER SOURCES

PRE-MECHANISATION

Until the application of the coal-fired steam engine to ships in the early 19th century, oars or the wind were used to assist watercraft propulsion. Merchant ships predominantly used sail, but during periods when naval warfare depended on ships closing to ram or to fight hand-to-hand, galley were preferred for their manoeuvrability and speed. The Greek navies that fought in the Peloponnesian War used triremes, as did the Romans at the Battle of Actium. The development of naval gunnery from the 16th century onward meant that manoeuvrability took second place to broadside weight; this led to the dominance of the sail-powered warship over the following three centuries.

In modern times, human propulsion is found mainly on small boats or as auxiliary propulsion on sailboats. Human propulsion includes the push pole, rowing, and pedals. Propulsion by sail generally consists of a sail hoisted on an erect mast, supported by stays, and controlled by lines made of rope. Sails were the dominant form of commercial propulsion until the late nineteenth century, and continued to be used well into the twentieth century on routes where wind was assured and coal was not available, such as in the South American nitrate trade. Sails are now generally used for recreation and racing, although experimental sail systems, such as the kites/royals, turbosails, rotorsails, wingsails, windmills and SkySails's own kite buoy-system have been used on larger modern vessels for fuel savings.

RECIPROCATING STEAM ENGINES

The development of piston-engined steamships was a complex process. Early steamships were fueled by wood, later ones by coal or fuel oil. Early ships used stern or side paddle wheels, while later ones used screw propellers. The first commercial success accrued to Robert Fulton's North River Steamboat (often called Clermont) in the US in 1807, followed in Europe by the 45-foot Comet of 1812. Steam propulsion progressed considerably over the rest of the 19th century. Notable developments included the steam surface condenser, which eliminated the use of sea water in the ship's boilers. This permitted higher steam pressures, and thus the use of higher efficiency multiple expansion (compound) engines. As the means of transmitting the engine's power, paddle wheels gave way to more efficient screw propellers.

STEAM TURBINES

Steam turbines were fueled by coal or, later, fuel oil or nuclear power. The marine steam turbine developed by Sir Charles Algernon Parsons raised the power-to-weight ratio. He achieved publicity by demonstrating it unofficially in the 100-foot Turbinia at the Spithead Naval Review in 1897. This facilitated a generation of high-speed liners in the first half of the 20th century, and rendered the reciprocating steam engine obsolete; first in warships, and later in merchant vessels.

In the early 20th century, heavy fuel oil came into more general use and began to replace coal as the fuel of choice in steamships. Its great advantages were convenience, reduced manpower by removal of the need for trimmers and stokers, and reduced space needed for fuel bunkers.

In the second half of the 20th century, rising fuel costs almost led to the demise of the steam turbine. Most new ships since around 1960 have been built with diesel engines. The last major passenger ship built with steam turbines was the Fairsky, launched in 1984. Similarly, many steam ships were re-engined to improve fuel efficiency. One high profile example was the 1968 built Queen Elizabeth 2 which had her steam turbines replaced with a diesel-electric propulsion plant in 1986. Most new-build ships with steam turbines are specialist vessels such as nuclear-powered vessels, and certain merchant vessels (notably Liquefied Natural Gas (LNG) and coal carriers) where the cargo can be used as bunker fuel.

LNG CARRIERS

New LNG carriers (a high growth area of shipping) continue to be built with steam turbines. The natural gas is stored in a liquid state in cryogenic vessels aboard these ships, and a small amount of 'boil off' gas is needed to maintain the pressure and temperature inside the vessels within operating limits. The 'boil off' gas provides the fuel for the ship's boilers, which provide steam for the turbines, the simplest way to deal with the gas. Technology to operate internal combustion engines (modified marine two-stroke diesel engines) on this gas has improved, however, so such engines are starting to appear in LNG carriers; with their greater thermal efficiency, less gas is burnt. Developments have also been made in the process of re-liquefying 'boil off' gas, letting it be returned to the cryogenic tanks.

The financial returns on LNG are potentially greater than the cost of the marine-grade fuel oil burnt in conventional diesel engines, so the re-liquefaction process is starting to be used on diesel engine propelled LNG carriers. Another factor driving the change from turbines to diesel engines for LNG carriers is the shortage of steam turbine qualified seagoing engineers. With the lack of turbine powered ships in other shipping sectors, and the rapid rise in size of the worldwide LNG fleet, not enough have been trained to meet the demand. It may be that the days are numbered for marine steam turbine propulsion systems, even though all but sixteen of the orders for new LNG carriers at the end of 2004 were for steam turbine propelled ships.

NUCLEAR-POWERED STEAM TURBINES

In these vessels, the nuclear reactor heats water to create steam to drive the turbines. Due to low prices of diesel oil, nuclear propulsion is rare except in some Navy and specialist vessels such as icebreakers. In large aircraft carriers, the space formerly used for ship's bunkerage could be used instead to bunker aviation fuel. In submarines, the ability to run submerged at high speed and in relative quiet for long periods holds obvious advantages. A few cruisers have also employed nuclear power; as of 2006, the only ones remaining in service are the Russian Kirov class. An example of a non-military ship with nuclear marine propulsion is the Arktika class icebreaker with 75,000 shaft horsepower (55,930 kW). Commercial experiments such as the NS Savannah have so far proved uneconomical compared with conventional propulsion.

In recent times, there is some renewed interest in commercial nuclear shipping. Nuclear-powered cargo ships could lower costs associated with carbon dioxide emissions and travel at higher cruise speeds than conventional diesel powered vessels.

RECIPROCATING DIESEL ENGINES

Most modern ships use a reciprocating diesel engine as their prime mover, due to their operating simplicity, robustness and fuel economy compared to most other prime mover mechanisms. The rotating crankshaft can be directly coupled to the propeller with slow speed engines, via a reduction gearbox for medium and high speed engines, or via an alternator and electric motor in diesel-electric vessels. The rotation of the crankshaft is connected to the camshaft or a hydraulic pump on an intelligent diesel.

The reciprocating marine diesel engine first came into use in 1903 when the diesel electric rivertanker Vandal was put into service by Branobel. Diesel engines soon offered greater efficiency than the steam turbine, but for many years had an inferior power-to-space ratio. The advent of turbocharging however hastened their adoption, by permitting greater power densities.

RECIPROCATING DIESEL ENGINES

Most modern ships use a reciprocating diesel engine as their prime mover, due to their operating simplicity, robustness and fuel economy compared to most other prime mover mechanisms. The rotating crankshaft can be directly coupled to the propeller with slow speed engines, via a reduction gearbox for medium and high speed engines, or via an alternator and electric motor in diesel-electric vessels. The rotation of the crankshaft is connected to the camshaft or a hydraulic pump on an intelligent diesel.

The reciprocating marine diesel engine first came into use in 1903 when the diesel electric rivertanker Vandal was put into service by Branobel. Diesel engines soon offered greater efficiency than the steam turbine, but for many years had an inferior power-to-space ratio. The advent of turbocharging however hastened their adoption, by permitting greater power densities.

Diesel engines today are broadly classified according to

- Their operating cycle: two-stroke engine or four-stroke engine
- Their construction: crosshead, trunk, or opposed piston
- Their speed
  - Slow speed: any engine with a maximum operating speed up to 300 revolutions per minute (rpm), although most large two-stroke slow speed diesel engines operate below 120 rpm. Some very long stroke engines have a maximum speed of around 80 rpm. The largest, most powerful engines in the world are slow speed, two stroke, crosshead diesels.
  - Medium speed: any engine with a maximum operating speed in the range 300-900 rpm. Many modern four-stroke medium speed diesel engines have a maximum operating speed of around 500 rpm.
  - High speed: any engine with a maximum operating speed above 900 rpm.

Most modern larger merchant ships use either slow speed, two stroke, crosshead engines, or medium speed, four stroke, trunk engines. Some smaller vessels may use high speed diesel engines.

The size of the different types of engines is an important factor in selecting what will be installed in a new ship. Slow speed two-stroke engines are much taller, but the footprint required, is smaller than that needed for equivalently rated four-stroke medium speed diesel engines. As space above the waterline is at a premium in passenger ships and ferries (especially ones with a car deck), these ships tend to use multiple medium speed engines resulting in a longer, lower engine room than that needed for two-stroke diesel engines. Multiple engine installations also give redundancy in the event of mechanical failure of one or more engines, and the potential for greater efficiency over a wider range of operating conditions.

As modern ships' propellers are at their most efficient at the operating speed of most slow speed diesel engines, ships with these engines do not generally need gearboxes. Usually such propulsion systems consist of either one or two propeller shafts each with its own direct drive engine. Ships propelled by medium or high speed diesel engines may have one or two (sometimes more) propellers, commonly with one or more engines driving each propeller shaft through a gearbox. Where more than one engine is geared to a single shaft, each engine will most likely drive through a clutch, allowing engines not being used to be disconnected from the gearbox while others keep running. This arrangement lets maintenance be carried out while under way, even far from port.

LNG ENGINES

Dual fuel engines are fueled by either marine grade diesel, heavy fuel oil, or liquefied natural gas (LNG). Having multiple fuel options will allow vessels to transit without relying on one type of fuel. Studies show that LNG is the most efficient of fuels although limited access to LNG fueling stations limits the production of such engines. Vessels providing services in the LNG industry have been retrofitted with dual-fuel engines and have been proved to be extremely effective. Benefits of dual-fuel engines include fuel and operational flexibility, high efficiency, low emissions, and operational cost advantages. Liquefied natural gas engines offer the marine transportation industry with an environmentally friendly alternative to provide power to vessels. In 2010 STX Finland and Viking Line signed an agreement to begin construction on what would be the largest environmentally friendly cruise ferry. Construction of NB 1376 will be completed in 2013. According to Viking Line, vessel NB 1376 will primarily be fueled by liquefied natural gas. The cruise ferry will have an emission reduction comparison to diesel-fuelled engines of approximately 90%. Vessel NB 1376 nitrogen oxide emissions will be almost zero and sulphur oxide emissions will be at least 80% below the International Maritime Organization's (IMO) standards. Company profits from tax cuts and operational cost advantages has led to the gradual growth of LNG fuel use in engines.

GAS TURBINES

Many warships built since the 1960s have used gas turbines for propulsion, as have a few passenger ships, like the jetfoil. Gas turbines are commonly used in combination with other types of engine. Most recently, the Queen Mary 2 has had gas turbines installed in addition to diesel engines. Because of their poor thermal efficiency at low power (cruising) output, it is common for ships using them to have diesel engines for cruising, with gas turbines reserved for when higher speeds are needed however, in the case of passenger ships the main reason for installing gas turbines has been to allow a reduction of emissions in sensitive environmental areas or while in port. Some warships, and a few modern cruise ships have also used steam turbines to improve the efficiency of their gas turbines in a combined cycle, where waste heat from a gas turbine exhaust is utilized to boil water and create steam for driving a steam turbine. In such combined cycles, thermal efficiency can be the same or slightly greater than that of diesel engines alone; however, the grade of fuel needed for these gas turbines is far more costly than that needed for the diesel engines, so the running costs are still higher.

SCREWS

Marine propellers are also known as "screws". There are many variations of marine screw systems, including twin, contra-rotating, controllable-pitch, and nozzle-style screws. While smaller vessels tend to have a single screw, even very large ships such as tankers, container ships and bulk carriers may have single screws for reasons of fuel efficiency. Other vessels may have twin, triple or quadruple screws. Power is transmitted from the engine to the screw by way of a propeller shaft, which may or may not be connected to a gearbox.

PADDLE WHEELS

The paddle wheel is a large wheel, generally built of a steel framework, upon the outer edge of which are fitted numerous paddle blades (called floats or buckets). The bottom quarter or so of the wheel travels underwater. Rotation of the paddle wheel produces thrust, forward or backward as required. More advanced paddle wheel designs have featured feathering methods that keep each paddle blade oriented closer to vertical while it is in the water; this increases efficiency. The upper part of a paddle wheel is normally enclosed in a paddlebox to minimise splashing.

Paddle wheels have been superseded by screws, which are a much more efficient form of propulsion. Nevertheless, paddle wheels have two advantages over screws, making them suitable for vessels in shallow rivers and constrained waters: first, they are less likely to be clogged by obstacles and debris; and secondly, when contra-rotating, they allow the vessel to spin around its own vertical axis. Some vessels had a single screw in addition to two paddle wheels, to gain the advantages of both types of propulsion.

SAILING

The purpose of sails is to use wind energy to propel the vessel, sled, board, vehicle or rotor.

WATER CATERPILLAR

An early uncommon means of boat propulsion was the water caterpillar. This moved a series of paddles on chains along the bottom of the boat to propel it over the water and preceded the development of tracked vehicles. The first water caterpillar was developed by Desblancs in 1782 and propelled by a steam engine. In the United States the first water caterpillar was patented in 1839 by William Leavenworth of New York.

BUOYANCY

Underwater gliders convert buoyancy to thrust, using wings, or more recently hull shape (SeaExplorer Glider). Buoyancy is made alternatively negative and positive, generating tooth-saw profiles.

PROPELLER (MARINE)

A propeller is a type of fan that transmits power by converting rotational motion into thrust. A pressure difference is produced between the forward and rear surfaces of the airfoil-shaped blade, and a fluid (such as air or water) is accelerated behind the blade. Propeller dynamics can be modelled by both Bernoulli's principle and Newton's third law. A marine propeller is sometimes colloquially known as a screw propeller or screw.

HISTORY

The principle employed in using a screw propeller is used in sculling. It is part of the skill of propelling a Venetian gondola but was used in a less refined way in other parts of Europe and probably elsewhere. For example, propelling a canoe with a single paddle using a "pitch stroke" or side slipping a canoe with a "scull" involves a similar technique. In China, sculling, called "lu", was also used by the 3rd century AD. In sculling, a single blade is moved through an arc, from side to side taking care to keep presenting the blade to the water at the effective angle. The innovation introduced with the screw propeller was the extension of that arc through more than 360° by attaching the blade to a rotating shaft. Propellers can have a single blade, but in practice there are nearly always more than one so as to balance the forces involved.

The origin of the screw propeller starts with Archimedes, who used a screw to lift water for irrigation and bailing boats, so famously that it became known as Archimedes' screw. It was probably an application of spiral movement in space (spirals were a special study of Archimedes) to a hollow segmented water-wheel used for irrigation by Egyptians for centuries. Leonardo da Vinci adopted the principle to drive his theoretical helicopter, sketches of which involved a large canvas screw overhead.

In 1784, J. P. Paucton proposed a gyrocopter-like aircraft using similar screws for both lift and propulsion. Robert Wilson of Dunbar, Scotland, demonstrated propulsion of a boat using a screw propeller in 1827, and gave further demonstrations in April 1828 in the Firth of Forth near to Leith. John Patch, a mariner in Yarmouth, Nova Scotia developed a two-bladed, fan-shaped propeller in 1832 and publicly demonstrated it in 1833, propelling a row boat across Yarmouth Harbour and a small coastal schooner at Saint John, New Brunswick, but his patent application in the United States was rejected until 1849 because he was not an American citizen. His efficient design drew praise in American scientific circles but by this time there were multiple competing versions of the marine propeller.

In 1835 Francis Pettit Smith discovered a new way of building propellers. Up to that time, propellers were literally screws, of considerable length. But during the testing of a boat propelled by one, the screw snapped off, leaving a fragment shaped much like a modern boat propeller. The boat moved faster with the broken propeller. At about the same time, Frédéric Sauvage and John

Ericsson applied for patents on vaguely similar, although less efficient shortened-screw propellers, leading to an apparently permanent controversy as to who the official inventor is among those three men. Ericsson became widely famous when he built the Monitor, an armoured battleship that in 1862 fought the Confederate States' Virginia in an American Civil War sea battle. The superiority of screw against paddles was taken up by navies. Trials with Smith's SS Archimedes, the first steam driven screw, led to the famous tug-of-war competition in 1845 between the screw-driven HMS Rattler and the paddle steamer HMS Alecto; the former pulling the latter backward at 2.5 knots (4.6 km/h). In the second half of the nineteenth century, several theories were developed. The momentum theory or disk actuator theory-a theory describing a mathematical model of an ideal propeller-was developed by W.J.M. Rankine (1865), Alfred George Greenhill (1888) and R.E. Froude (1889). The propeller is modelled as an infinitely thin disc, inducing a constant velocity along the axis of rotation. This disc creates a flow around the propeller. Under certain mathematical premises of the fluid, there can be extracted a mathematical connection between power, radius of the propeller, torque and induced velocity. Friction is not included.

The blade element theory (BET) is a mathematical process originally designed by William Froude (1878), David W. Taylor (1893) and Stefan Drzewiecki to determine the behaviour of propellers. It involves breaking an airfoil down into several small parts then determining the forces on them. These forces are then converted into accelerations, which can be integrated into velocities and positions. In 2009, Mike Richard John Smith filed a Canadian Patent application (Canadian Patent number: 2675044) indicating the screw type propeller is inefficient due to the screw type blade orientation of the propeller which when in operation for forward movement substantially moves the fluid (for a boat the fluid is water)sidewardly, rotationally, and rearwardly wherein moving the fluid sidewardly does not move the boat forward, thus contributing to the inefficiency of the screw type propeller.

Marine propeller nomenclature

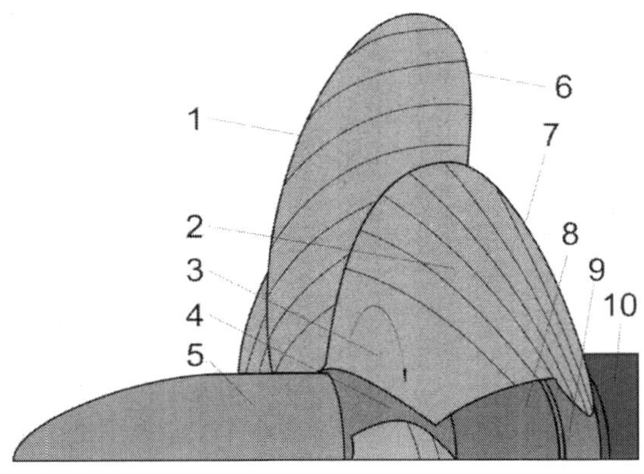

1. Trailing edge
2. Face
3. Fillet area
4. Hub or Boss
5. Hub or Boss Cap
6. Leading edge
7. Back
8. Propeller shaft
9. Stern tube bearing
10. Stern tube

A propeller is the most common propulsor on ships, imparting momentum to a fluid which causes a force to act on the ship. The ideal efficiency of any size propeller (free-tip) is that of an actuator disc in an ideal fluid. An actual marine propeller is made up of sections of helicoidal surfaces which act together 'screwing' through the water (hence the common reference to marine propellers as "screws"). Three, four, or five blades are most common in marine propellers, although designs which are intended to operate at reduced noise will have more blades. The blades are attached to a boss (hub), which should be as small as the needs of strength allow - with fixed pitch propellers the blades and boss are usually a single casting.

An alternative design is the controllable pitch propeller (CPP, or CRP for controllable-reversible pitch), where the blades are rotated normally to the drive shaft by additional machinery - usually hydraulics - at the hub and control linkages running down the shaft. This allows the drive machinery to operate at a constant speed while the propeller loading is changed to match operating conditions. It also eliminates the need for a reversing gear and allows for more rapid change to thrust, as the revolutions are constant. This type of propeller is most common on ships such as tugs where there can be enormous differences in propeller loading when towing compared to running free, a change which could cause conventional propellers to lock up as insufficient torque is generated. The downsides of a CPP/CRP include: the large hub which decreases the torque required to cause cavitation, the mechanical complexity which limits transmission power and the extra blade shaping requirements forced upon the propeller designer.

For smaller motors there are self-pitching propellers. The blades freely move through an entire circle on an axis at right angles to the shaft. This allows hydrodynamic and centrifugal forces to 'set' the angle the blades reach and so the pitch of the propeller. A propeller that turns clockwise to produce forward thrust, when viewed from aft, is called right-handed. One that turns anticlockwise is said to be left-handed. Larger vessels often have twin screws to reduce heeling torque, counter-rotating propellers, the starboard screw is usually right-handed and the port left-handed, this is called outward turning. The opposite case is called inward turning. Another possibility is contra-rotating propellers, where two propellers rotate in opposing directions on a single shaft, or on separate shafts on nearly the same axis. One example of the latter is the CRP Azipod by the ABB Group. Contra-rotating propellers offer increased efficiency by capturing the energy lost in the tangential velocities imparted to the fluid by the forward propeller (known as "propeller swirl"). The flow field behind the aft propeller of a contra-rotating set has very little "swirl", and this reduction in energy loss is seen as an increased efficiency of the aft propeller.

ADDITIONAL DESIGNS

An azimuthing propeller is a propeller that turns around the vertical axis. The individual airfoil-shaped blades turn as the propeller moves so that they are always generating lift in the vessel's direction of movement. This type of propeller can reverse or change its direction of thrust very quickly,

HISTORY OF SHIP AND SUBMARINE SCREW PROPELLERS

James Watt of Scotland is generally credited with applying the first screw propeller to an engine at his Birmingham works, an early steam engine, beginning the use of an hydrodynamic screw for propulsion. Mechanical ship propulsion began with the steam ship. The first successful ship of this type is a matter of debate; candidate inventors of the 18th century include William Symington, the Marquis de Jouffroy, John Fitch and Robert Fulton, however William Symington's ship the Charlotte Dundas is regarded as the world's "first practical steamboat". Paddlewheels as the main motive source became standard on these early vessels (see Paddle steamer). Robert Fulton had tested, and rejected, the screw propeller.

The screw (as opposed to paddlewheels) was introduced in the latter half of the 18th century. David Bushnell's invention of the submarine (Turtle) in 1775 used hand-powered screws for vertical and horizontal propulsion. The Bohemian (Austrian Empire) engineer Josef Ressel designed and patented the first practicable screw propeller in 1827. Francis Pettit Smith tested a similar one in 1836. In 1839, John Ericsson introduced practical screw propulsion into the United States. Mixed paddle and propeller designs were still being used at this time (vide the 1858 Great Eastern).

The screw propeller replaced the paddles owing to its greater efficiency, compactness, less complex power transmission system, and reduced susceptibility to damage (especially in battle) Initial designs owed much to the ordinary screw from which their name derived - early propellers consisted of only two blades and matched in profile the length of a single screw rotation. This design was common, but inventors endlessly experimented with different profiles and greater numbers of blades. The propeller screw design stabilized by the 1880s. In the early days of steam power for ships, when both paddle wheels and screws were in use, ships were often characterized by their type of propellers, leading to terms like screw steamer or screw sloop. Propellers are referred to as "lift" devices, while paddles are "drag" devices.

MARINE PROPELLER CAVITATION

Cavitation is the formation of vapor bubbles in water near a moving propeller blade in regions of low pressure due to Bernoulli's principle. It can occur if an attempt is made to transmit too much power through the screw, or if the propeller is operating at a very high speed. Cavitation can waste power, create vibration and wear, and cause damage to the propeller. It can occur in many ways on a propeller. The two most common types of propeller cavitation are suction side surface cavitation and tip vortex cavitation.

Suction side surface cavitation forms when the propeller is operating at high rotational speeds or under heavy load (high blade lift coefficient). The pressure on the upstream surface of the blade (the "suction side") can drop below the vapor pressure of the water, resulting in the formation of a vapor pocket. Under such conditions, the change in pressure between the downstream surface of the blade (the "pressure side") and the suction side is limited, and eventually reduced as the extent of cavitation is increased. When most of the blade surface is covered by cavitation, the pressure difference between the pressure side and suction side of the blade drops considerably, as does the thrust produced by the propeller. This condition is called "thrust breakdown". Operating the propeller under these conditions wastes energy, generates considerable noise, and as the vapor bubbles collapse it rapidly erodes the screw's surface due to localized shock waves against the blade surface.

Tip vortex cavitation is caused by the extremely low pressures formed at the core of the tip vortex. The tip vortex is caused by fluid wrapping around the tip of the propeller; from the pressure side to the suction side. This video demonstrates tip vortex cavitation. Tip vortex cavitation typically occurs before suction side surface cavitation and is less damaging to the blade, since this type of cavitation doesn't collapse on the blade, but some distance downstream.

Cavitation can be used as an advantage in design of very high performance propellers, in form of the supercavitating propeller. In this case, the blade section is designed such that the pressure side stays wetted while the suction side is completely covered by cavitation vapor. Because the suction side is covered with vapor instead of water it encounters very low viscous friction, making the supercavitating (SC) propeller comparably efficient at high speed. The shaping of SC blade sections however, make it inefficient at low speeds, when the suction side of the blade is wetted. (See also fluid dynamics). A similar, but quite separate issue, is ventilation, which occurs when a propeller operating near the surface draws air into the blades, causing a similar loss of power and shaft vibration, but without the related potential blade surface damage caused by cavitation. Both effects can be mitigated by increasing the submerged depth of the propeller: cavitation is reduced because the hydrostatic pressure increases the margin to the vapor pressure, and ventilation because it is further from surface waves and other air pockets that might be drawn into the slipstream.

The blade profile of propellers designed to be operate in a ventilated condition is often not of an aerofoil section and is a blunt ended taper instead. These are often known as "chopper" type propellers.

TYPES OF MARINE PROPELLERS

Controllable pitch propeller

One type of marine propeller is the controllable pitch propeller. This propeller has several advantages with ships. These advantages include: the least drag depending on the speed used, the ability to move the sea vessel backwards, and the ability to use the "vane"-stance, which gives the least water resistance when not using the propeller (e.g. when the sails are used instead).

Skewback propeller

An advanced type of propeller used on German Type 212 submarines is called a skewback propeller. As in the scimitar blades used on some aircraft, the blade tips of a skewback propeller are swept back against the direction of rotation. In addition, the blades are tilted rearward along the longitudinal axis, giving the propeller an overall cup-shaped appearance. This design preserves thrust efficiency while reducing cavitation, and thus makes for a quiet, stealthy design.

Modular propeller

A modular propeller provides more control over the boats performance. There is no need to change an entire prop, when there is an opportunity to only change the pitch or the damaged blades. Being able to adjust pitch will allow for boaters to have better performance while in different altitudes, water sports, and/or cruising.

PROTECTION OF SMALL ENGINES

For smaller engines, such as outboards, where the propeller is exposed to the risk of collision with heavy objects, the propeller often includes a device which is designed to fail when over loaded; the device or the whole propeller is sacrificed so that the more expensive transmission and engine are not damaged.

Typically in smaller (less than 10 hp or 7.5 kW) and older engines, a narrow shear pin through the drive shaft and propeller hub transmits the power of the engine at normal loads. The pin is designed to shear when the propeller is put under a load that could damage the engine. After the pin is sheared the engine is unable to provide propulsive power to the boat until an undamaged shear pin is fitted. Note that some shear pins used to have shear grooves machined into them. Nowadays the grooves tend to be omitted. The result of this oversight is that the torque required to shear the pin rises as the cutting edges of the propeller bushing and shaft become blunted. Eventually the gears will strip instead.

In larger and more modern engines, a rubber bushing transmits the torque of the drive shaft to the propeller's hub. Under a damaging load the friction of the bushing in the hub is overcome and the rotating propeller slips on the shaft preventing overloading of the engine's components. After such an event the rubber bushing itself may be damaged. If so, it may continue to transmit reduced power at low revolutions but may provide no power, due to reduced friction, at high revolutions. Also the rubber bushing may perish over time leading to its failure under loads below its designed failure load.

Whether a rubber bushing can be replaced or repaired depends upon the propeller; some cannot. Some can but need special equipment to insert the oversized bushing for an interference fit. Others can be replaced easily.

The "special equipment" usually consists of a tapered funnel, some kind of press and rubber lubricant (soap). Often the bushing can be drawn into place with nothing more complex than a couple of nuts, washers and "allscrew" (threaded bar). If one does not have access to a lathe an improvised funnel can be made from steel tube and car body filler (as the filler is only subject to compressive forces it is able to do a good job) A more serious problem with this type of propeller is a "frozen-on" spline bushing which makes propeller removal impossible. In such cases the propeller has to be heated in order to deliberately destroy the rubber insert. Once the propeller proper is removed, the splined tube can be cut away with a grinder. A new spline bushing is of course required. To prevent the problem recurring the splines can be coated with anti-seize anti-corrosion compound.

In some modern propellers, a hard polymer insert called a drive sleeve replaces the rubber bushing. The splined or other non-circular cross section of the sleeve inserted between the shaft and propeller hub transmits the engine torque to the propeller, rather than friction. The polymer is weaker than the components of the propeller and engine so it fails before they do when the propeller is overloaded. This fails completely under excessive load but can easily be replaced.

FORCES ON SAILS

Forces on sails are primarily due to movement of air near and relative to the sails. Understanding the forces on sails is important for the design and operation of the sails and whatever they are moving, sailboats, ice boats, sailboards, land sailing vehicles or windmill sail rotors.

Aerodynamic forces occur along the entire surface of the sails, but can be summed into one net force vector. Net aerodynamic force may be decomposed with respect to a boat's course over water into components acting in six degrees of freedom. Two components with respect to wind direction can also be resolved: drag, which is the component directed down wind, and lift, which is the component normal to the wind and perpendicular to drag. This analysis is important to boat design, operation, balance, stability, seakindliness and seaworthiness.

Briefly, when the sail is oriented at a right angle to the wind, as in a boat sailing downwind, the aerodynamic force, by definition, is almost entirely derived from the drag component - the wind "pushes" the sail along in the direction of the wind.

When the sail is arranged across or into the wind the sail acts as an airfoil. The resulting surface force includes a lift or pressure component normal to the sail surface as well as a tangential drag stress component. Various mathematical models explain the extent and direction of the aerodynamic force on the sail(s). By the law of conservation of momentum, the wind moves the sail as the sail redirects the air backwards .

OVERVIEW

The analysis of the forces on sails takes into account the theoretical location of the propulsive force or centre of effort, the direction of the force, and the intensity and distribution of the pressure related surface force and/or lift. The fluid mechanics and aerodynamics airflow calculations for a boat are more complex than for a rigid winged aircraft.

Structural analysis also is involved in modern optimal sail design and manufacture. Aeroelasticity models, combining computational fluid dynamics and structural analysis, are at the frontiers of sail study and design. However, turbulence and detachment of the boundary layer are not yet fully understood. Computational limitations persist.

The theoretical results are corrected by reality. So, wind tunnel scale model and full scale testing of sails are required for optimum sail design, function and trim.

Some complexities of boat sails:

- The wind is not constant.
- The boat is not traveling in uniform velocity.
- There may be a mast in front of the sail, disturbing the airflow, although this may be mitigated by profiling it.
- A mast is not infinitely stiff.
- The boat profile and position influence the airflow.
- A sail is usually made of thin and deformable fabric.
- The air is viscous, causing losses by friction.
- The flow of the air varies from slow to fast and turbulent to laminar.

Though some software algorithms attempt to model these complexities, the following assumptions make the analysis much simpler:

- the water more or less flat
- the wind more or less constant
- the sail is set and is not adjusted

CENTRE OF EFFORT

The point of origin of net aerodynamic force on sails is the centre of effort (or also centre of pressure). In a first approximate approach, the location of the centre of effort is the geometric centre of the sail. Filled with wind, the sail has a roughly spherical polygon shape and if the shape is stable, then the location of centre of effort is stable. The position of centre of effort will vary with sail plan, sail trim or airfoil profile, boat trim and point of sail.

DIRECTION OF FORCE ON SAILS

The net aerodynamic force on the sail is located quasi at the maximum draught intersecting the camber of the sail and passing through a plane intersecting the centre of effort, normal to the mast, quasi perpendicular to the chord of the sail (a straight line between the leading edge (luff) and the trailing edge (leech)).

Net aerodynamic force may be decomposed into the three translation directions with respect to a boat's course in a seaway: surge (forward/astern); sway (starboard/port, relevant to leeway); heave (up/down). The force terms of torque in the three rotation directions, roll (rotation about surge axis, relevant to heeling).

Pitch (rotation about sway axis), yaw (rotation about heave axis, relevant to broaching) may be also derived. The scalar values and direction of these components may be very dynamic and dependent on many variables on a boat and in a seaway including the point of sail.

The net force vector, F_T, *is resolved into components in relation to course in a seaway with:*

- F_R = the driving force directed along the course sailed and
- F_H = the heeling force perpendicular to the course and the mast.

The heeling force can be resolved as a function of heel angle, t, to:

- F_{lat} @ F_H „ $cos(t)$, the lateral or leeway force

and

- F_{vert} @ F_H „ $sin(t)$, the vertical or heave force.

Net aerodynamic sail force can also be resolved into two components with respect to wind direction: drag, which is the component directed down wind, and lift, which is the component normal to the freestream wind and perpendicular to drag. The generation of lift and drag, components of F_T, and their contribution to boat motion are discussed below.

PRESSURE ON THE SAIL

For the purposes of modern sail making and study, pressure distribution measurements are done in wind tunnel and full scale experiments as well as in computer models.

According to kinetic theory, at the microscopic level, air pressure is the result of collisions between perpetually moving air particles. Their energy, measured by temperature, determines their velocity. In still air, the average particle of air randomly moves around an imaginary fixed point in space, colliding with other particles without too much average movement away from this point. Wind is the particles moving in large numbers in the same direction. So, air pressure on a sail has two origins: temperature and the mechanical influence of wind.

Pitot tubes and other types of manometers are used in wind tunnel and full scale testing to measure the differences between local static pressures at various points on the sail and atmospheric pressure (static pressure in undisturbed flow). Results are graphed as pressure coefficients (static pressure difference over wind induced dynamic pressure) to obtain windward "pressure" to leeward "suction" distribution curves along the mast/sail's chord.

ROLE OF ATMOSPHERIC PRESSURE

There are fewer air particles at high altitude. Collisions between slower, colder particles are less violent and less frequent. Thus, there is less pressure. At sea level there are more particles with more energy, resulting in more frequent and violent collisions, or higher pressure.

Close to the sail, collisions occur between sail and air particles. These collisions generate a force on the sail at sea level of about 10 tonnes-force per square meter of sail (101325 Pa). If the pressures on each side of a sail are perfectly balanced, the sail does not move.

LIFT'S EFFECT ON THE SAIL

To study the effect of lift we can compare cases with and without lift. As an approximation of a gaff sail, take a sail that is rectangular and approximately vertical, with an area of 10 m^2 - 2.5 m of foot by 4 m of leech. The apparent wind is 8.3 m/s (about 30 km/h). The boat is presumed to have uniform velocity, no heel and no pitch and there are no waves. The density of air is set at: u = 1.2 kg/m^3.

SAILING IN STALLED FLOW

The boat is running downwind. The shape of the sail is approximated by a plane perpendicular to the apparent wind. The depression effect on the sail is second order, and therefore negligible. The remaining pressures are:

- on the windward side atmospheric pressure and wind pressure
- on the leeward side only the atmospheric pressure

Forces of atmospheric pressure cancel out. There remains only pressure generated by the wind.

SAILING IN ATTACHED FLOW

The boat is close hauled, with the sail set at, for example, 15° relative to the apparent wind. The camber of the sail creates a lift. In other words, the effect of depression on the leeward side comes into play. As air pressure forces cancel out, significant resulting forces are:

- on the windward side wind pressure
- on the leeward side wind depression

The only unknown to be determined is the drag coefficient. In a well trimmed sail the curve profile is close to optimal airfoil shape NACA 0012. A less well trimmed sail, perhaps of older technology, will have greater draft with more camber. The coefficient of aerodynamic lift will be higher but the sail will be less efficient with a lower lift/drag ratio (L/D). The sail profile may be similar to NACA 0015, NACA 0018.

For a given profile, there are tables which give the lift coefficient (Cz), which depends on several variables:

- Incidence angle of apparent wind to sail profile,
- The slope of lift of the sail, which depends on its Aspect ratio,
- The surface roughness and Reynolds number, which affect the flow of fluid (laminar, turbulent).

The coefficient is determined for a stable and uniform fluid, and a profile of infinite extension.

INFLUENCE OF APPARENT WIND

When a ship is moving, its velocity creates a relative wind. The sum of the true wind and the relative wind is called the apparent wind. If the ship moves upwind, the two winds are cumulative, and the apparent wind is larger than the actual wind. Downwind, the effect is reversed, winds are subtracted and the apparent wind is lower than the true wind.

INFLUENCE OF RIGGING TENSION ON LIFT PERFORMANCE

Trimming a sail involves two parameters:

- Angle of attack, or incidence. i.e. the angle of apparent wind to sail chord to create maximum lift, or a maximum L/D. This angle varies with sail height, which is aerodynamic twist.
- Airfoil profile which is composed of: 1. Camber of the sail as defined as the ratio of maximum draft depth to chord length and 2. Draft position.

Sail set and shape is generally flexible. When the sail is operating in lift, if a sail is not properly inflated and stretched, there are wrinkles on the sail.

These folds form a break in the profile. The air does not slip along the sail. The air streams come off the airfoil profile. Areas of recirculation or turbulent separation bubbles appear. These areas considerably diminish the performance of the sail.

The assumption of a non wrinkled profile will simplify sail analysis. A sail may be rigid where the canopy is composed of non stretchy fiber. Tightening a flat piece of such cloth inflated by the wind results in folds at the attachment points. To avoid wrinkles, the sail could be tightened harder. The tension can be considerable to eliminate all wrinkles. So, in the case of a taut rigid sail, the inflated shape is static, hollow and with its draft position immobile.

The more elastic sail deforms slightly to its locations of high stress on the material, thereby eliminating wrinkles. The sail is no longer flat. Consequently, the sail can take several forms. By varying the tension of the sail, it is more or less empty. It is possible to vary the shape of the sail without folds. The potential sail shapes are intrinsically linked to the cut of the sail.

So in the elastic case, there is a family of possible forms and draft depths and positions the sail may take. Sailmakers try to build rigidity into sails for a predictable working shape with a degree of advantageous resilience depending on the sail's type, application and range: racing, cruising, high, moderate or variable wind, etc. The airfoil profile of the sail changes depending on the sail trim. At a given incidence, the sail can take different forms. The shape depends on the rigging tensions such as on clew corner of the sail, the tack with Cunningham adjustment, the backstay, the outhaul, the halyards or the boom vang (kicking strap). These elements help determine the shape of the sail. More exactly, they can decide position of maximum draft along the camber of the sail.

Each profile represents an appropriate value of Cz (lift coefficient). The position of the draft along the chord with the most lift is about 40% of the foot from luff. The leeward side of a sail is close to the NACA series 0012 (NACA 0015, NACA 0018, etc.) within the possibilities of trimming. The position of the draft is not independent of the camber setting. These parameters are linked by the shape of sail. Modifying the camber modifies the position of the draft.

CAMBER

The curves of propulsive component of lift and heel versus the angle of attack vary with the camber of the sail, that is to say, the biggest draft depth relative to the chord of the sail. A sail with high camber has a higher aerodynamic coefficient and, potentially, a greater propulsive force. Though the heeling coefficient varies with draft depth in the same direction. So finding the optimal camber will be a compromise between achieving a large propulsive force and an acceptable list. ·

DRAFT POSITION

The curves of propulsive lift and heel as a function of the angle of attack also depend on the position of the draft's proximity to the luff. · .

INFLUENCE OF ASPECT RATIO AND SAIL PLANFORM ON INDUCED DRAG

Sails are not infinitely long. They have ends. For the mainsail:

- Boom
- Head.

The transfer of air molecules from the windward pressured side to the leeward depressed side around the edge the thin sail is very violent. This creates significant turbulence, loss of pressure difference and loss of propulsion. On the end of a wing this is manifest as wingtip vortex. On a Bermuda sail, foot and leech are two areas where this phenomenon exists. The drag of leech is included in drag in the usual lift curves. The sail airfoil profile is considered as infinite (i.e. no ends). But foot drag is calculated separately. This loss of efficiency of the sail at the foot is called Lift-induced drag.

INFLUENCE OF THE HEIGHT OF THE FOOT RELATIVE TO SEA LEVEL

The gap between the edge of the sail and the sea surface has a significant influence on performance of a Bermudan type sail. In effect it creates an additional trailing edge vortex.

The vortex would be nonexistent if the border were in contact with the sea. This vortex consumes extra energy and thus modifies the coefficients of lift and drag. The hole is not completely empty, as the sail is partially filled by the freeboard and superstructure of any sailboat.

For a height between the edge of the sail and the deck of the sailboat of 6% of the length of the mast, changes are:

- a 20% increase in the drag coefficient
- a 10% loss in the lift coefficient.

The crab claw sail may partially circumvent this problem by harnessing the delta-wing's vortex lift.

Shape of luff, leech, and foot

A sail hauled up has a three-dimensional shape. This form is chosen by the sailmaker. The 3D shape is different for the hauled up form compared to when empty of wind. This must be taken into account when cutting the sail. The general shape of a sail is a deformed polygon. The polygon is slightly distorted in the case of a Bermuda sail and heavily distorted in the case of a spinnaker. The shape of edges empty is different from shape of edges once the sail is hauled up. Convex empty can go to straight edge when the sail is hauled up.

Edges can be:

- convex
- concave
- straight

When the convex shape is not natural (except for a free edge in a spinnaker), the sail is equipped with battens to maintain this pronounced convex shape. Except for the spinnaker with a balloon shape, the variation of edge empty compared to straight line remains low, a few centimeters.

Once hauled up, an elliptical sail would be ideal. But as the sail is not rigid:

- You need a mast, which for reasons of technical feasibility, needs to be quite straight.
- Flexibility of the sail can bring other problems, which are better to fix at the expense of an ideal convex elliptic shape.

LEECH

On a Bermudan type sail the oval is the ideal (convex), but a concave shaped leech improves the twist at the top of the sail and prevents overpowering the top of the sail in the gusts, thereby improving the boat's stability. The concave leech makes sailing more tolerant and more neutral. A convex shape is an easy way to increase the sail area (roach). Marchaj discusses crescent shaped foils like a raked wing tip device as seen on various fish fins, Brazilian jungada sails, crab claw sails, and America's cup boat Stars and Stripes to reduce lift induced drag.

LUFF

Once hauled up, the edge must be parallel to the forestay or mast. Masts and spars are very often, except in windsurfing, jangada boats, and proas, straight. So, a straight luff is usually needed.

But the draft of the sail is normally closer to luff than foot. So to facilitate the implementation of draft of the sail when hauled up, the empty form of luff is convex. This convexity is called the luff curve. Sometimes rigging is complex and the mast is not straight. In this case, the shape of luff empty can be convex at bottom and concave at the top.

FOOT

Foot form has little importance, particularly on sails with a loose foot or free edge. Its shape is more motivated by aesthetic reasons. Often it is convex empty to be straight once hauled up. When the border is attached to a spar or boom a convex shape is preferred to facilitate formation of draft of the sail. On retractable booms, the shape of the edge of the border is chosen based on technical constraints associated with the reel than consideration of aerodynamics. A winglet as used on airplanes to minimise lift induced drag is so far not practically seen on sails.

RELATIONSHIP OF LIFT COEFFICIENT TO ANGLE OF INCIDENCE: POLAR DIAGRAM

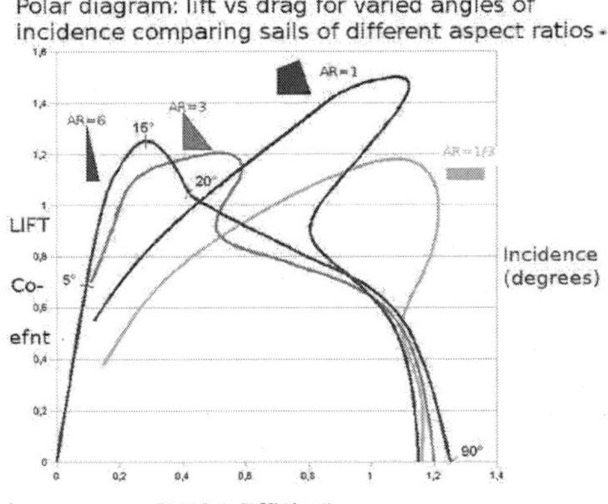

Polar curves showing the relationship between lift and drag for sails of aspect ratios 6, 3, 1 and 1/3 over varied incidence angles.

The aerodynamic coefficients of the sail vary with angle of attack (incidence of chord to apparent wind). The analysis on a polar diagram correlates to the respective lift and drag components of aerodynamic force:

- The component perpendicular to the apparent wind is called the lift coefficient;
- The component parallel to the apparent wind is called the drag coefficient.

Each incidence angle corresponds with a single lift-drag pair.

Summarising the behaviour of the sail at varying incidence:

- When the sail is loose, this is the equivalent to having no sail. Lift and drag from the sail are effectively null.
- When the sail is perpendicular to the wind, the air movement is turbulent. This is the case of no lift and maximum drag.
- These are the intermediate cases:
  - Sail loose to maximum lift: the flow is attached, i.e. there is an airfoil. There are no eddies (dead zones) created on the sail. It is noted in the case of a good well trimmed sail, maximum lift is greater than maximum drag.
  - Maximum lift to maximum dead zone: the wind does not stick properly to profile of the sail. Flow is less stable. Air becomes gradually lifted or taken off. This creates an area on leeward side, a dead zone where depressions form on the sail. At typical angle, dead zone has invaded the leeward side.
  - The dead zone to maximum drag: Dead zone has invaded the whole face on the leeward side, only on the windward side is

there an effect. Air in these high angles, is somewhat deviated from its trajectory. Air particles are just crashing on all surfaces of windward side. Force is almost constant, so the polar sailing describes an arc of a circle.

As the lift is more effective than drag in contributing to the advancement of ship, sail makers trying to increase the zone of lift, i.e. increase force of lift and angle of incidence. The task of a knowledgeable sailmaker is to decrease the size of the dead zone at high angles of incidence, i.e. to control the boundary layer.

INFLUENCE OF ALTITUDE: AERODYNAMIC TWIST AND SAIL TWIST

The rapid increase of the wind speed with altitude will increase both apparent wind speed and its angle of incidence to course sailed, (e). When using sails with lift, the sail must be twisted to have a consistent angle of incidence of sail with apparent wind, (d), along the leading edge (luff). This results in the lower sail chords being at smaller angles to the course sailed, (e - d), (see decomposition of forces diagram below) than the upper chords to compensate for the smaller e angle closer to the deck.

The air moves primarily in slices parallel to the ground or sea. While air density can be regarded as constant for our calculations of force, this is not the case for wind speed distribution. Wind speed will increase with altitude. At the sea surface, the difference of speed between air particles and water is zero. The wind speed increases strongly in the first ten meters. KW Ruggles gives a generally accepted formula for the relation of the wind speed with altitude:

$$U \, @ \, \frac{p^{\rceil}}{k} \, ln(\frac{z \; . \; z0}{z0})$$

With data collected by Rod Carr the parameters are:

- k = 0.42,
- z altitude in meters;
- z0 is an altitude that reflects the state of the sea, i.e. the wave height and speed:
  - 0.01 for 0-1 Beaufort;
  - 0.5 2-3 Beaufort
  - 5.0 to 4 Beaufort;
  - 20 5-6 Beaufort;
- $p^{\rceil}$= 0335 related to viscosity of air;
- U m / s.

In practice, the twist must be adjusted to optimize the performance of the sail. The primary means of control is the boom for a Bermuda mainsail. The more the boom is pulled down, the less twist. For the foresail, depending on the rig, twist is controlled by adjusting jib leech tension through sheet tension adjustments of: sheet angle with sheet block track (fair lead) position, jib halyard tension, jib Cunningham tension, or forestay tension.

INFLUENCE OF THE ROUGHNESS OF THE SAIL

As on a hull or wing, roughness plays a role on the performance of the sail. Small humps and hollows may have a stabilizing effect or facilitate stalls as when switching from laminar to turbulent flow. They also influence friction losses.

This area is the subject of research in real and wind tunnel conditions. It is currently not simulated numerically. It appears that at high Reynolds number, well chosen roughness prolongs the laminar mode incidence a few degrees more.

INFLUENCE OF THE REYNOLDS NUMBER

The Reynolds number is a measure of the ratio of inertial forces to viscous forces in moving fluids. It also indicates degrees of laminar or turbulent flow. Laminar flow occurs at low Reynolds numbers, where viscous forces are dominant, and is characterised by smooth, constant fluid motion. Turbulent flow occurs at high Reynolds numbers and is dominated by inertial forces.

The stronger the wind, the more the air particles tend to continue moving in a straight line, so are less likely to stick to the wing, making the transition to turbulent mode nearer. The higher the Reynolds number the better the performance of the sail (within other optimal parameters.)

The lift force formula

$$F \;@\; \frac{1}{2} \;„\; u \;„\; S \;„\; C \;„\; V^2$$

is practical and easy to use. The aerodynamic lift coefficient, C, depends on wind speed, V, and surface characteristics. The lift coefficient depends on Reynolds number as shown in the tables and polar diagrams. The Reynolds number is defined by

$$\mathrm{Re} \;@\; \frac{UL}{q}.$$

The Reynolds number depends on wind speed, U, and length, L, travelled by the air (characteristic chord length) and kinematic viscosity, q. But the influence of the Reynolds number is second order relative to other factors. The performance of the sail changes very little for a variation of the Reynolds number. The influence of very low Reynolds number is included within the tables (or chart) by plotting the lift coefficient (or drag) for several values of the Reynolds number (usually three values).

Increasing the incidence or the maximum lift coefficient by good choice of the Reynolds number is very interesting but secondary. The Reynolds number depends only on three parameters: speed, viscosity and length:

Viscosity is a physical constant, it is not an input variable for optimisation. Wind speed is a variable of optimisation. It is obvious that we look for the highest possible wind speed on the sail for sailing maximum force much more than for reasons of Reynolds number. This parameter has already been optimised.

The sail is inherently inelastic and of fixed size. So, the characteristic length is fixed for a given sail. Length optimisation is the responsibility of the naval architect, except for sail changes by the sailor. Performance tuning of the sails by varying the characteristic length of the Reynolds number is masked by the optimisation of other parameters, such as looking for better sailing performance by adjusting the weight of the sails. The weight of the sail is an important point for the balance of the ship. Just a little more weight in the higher part of the sail may create a major change affecting the balance of the ship. Or, for high winds, the sail fabric must resist tearing, so be heavy. The sailor is looking for a set of sails adapted to each range of wind speeds for reasons of weight more than for reasons of Reynolds number: jib, storm sail, main sail, spinnaker, light genoa, heavy genoa, etc. Each wind speed has its sail. Higher winds tend to force small characteristic lengths. The choice of the shape of the sails and therefore the characteristic length is guided by other criteria more important than the Reynolds number. The price of a sail is very high and therefore, limits the number of sails.

The coefficients of lift and drag, including the influence of the Reynolds number, are calculated by solving the equations of physics governing the flow of air over a wing using computed simulation models. The results found are well correlated with reality, less than 3% error.

LIFT/DRAG RATIO AND POWER

Polar curves of lift versus drag initially have a high slope. This is very well explained by the theory of thin profiles. The initial constant drag and lift slope becomes more horizontal, as maximum lift is approached. Then at higher angles of incidence a dead zone appears, reducing the effectiveness of the sail. The goal of the sailor is to set the sail in the incidence angle where the pressure is maximum. Considering the proper tuning for a Bermuda-rigged boat, it is rare to set a sail with theoretical optimum L/D. The apparent wind is not constant for two reasons: wind and sea. The wind itself is not constant, or even simply variant.

There are swings in the wind, there are gusts of wind and wind shifts. Even assuming constant wind, the boat can be raised with the swell or wave, the top of the sail finding faster winds, or in the troughs there is less wind. Up or down a wave the boat pitches, that is to say, the top of the sail is propelled forward and back constantly changing the apparent wind speed, relative to the sail. The apparent wind changes all the time and very quickly. It is often impossible to adapt to sea conditions with correspondeningly fast adjustments of the sails. Therefore, it is impossible to be at the theoretical optimum. This is not necessarily a disadvantage as the "pumping" phenomenon of abrupt changes in incidence has been shown to increase lift beyond the steady flow situation. Nevertheless, setting to the maximum optimum may prove quickly disastrous for a small change in wind. It is best to find an optimum setting more tolerant to changing conditions of apparent wind, state of equipment and weather.

The important parameter influencing the type of sail trim is the shape of the hull. The hull shape is elongated to provide a minimum of resistance to progress. We need to consider effects of wind on direction of the hull tilt: forward (pitch) or heel (roll). Downwind, sailing thrust is oriented in the direction of travel so will result in a forward pitch. Maximizing sail area may be important as the heeling force is minimal.

The situation changes if part of the force is perpendicular to the vessel. For the same force as sailing downwind, the force perpendicular to the vessel may result in a substantial heel. Under heavy list, the top of the sail does not take advantage of stronger winds at altitude, where the wind can give maximum energy to sail and boat. The heel phenomenon is much more sensitive than sail induced pitching. Accordingly, to minimize the list, the type of setting will be different close to the wind versus downwind: Close hauled, the setting is for L/D. When sailing downwind, the setting is for power.

LIFT/DRAG. UPWIND SAIL CUT AND TRIM

In the example of upwind sailing, the apparent wind, with incidence, d, to the sail chord, is at an angle, e, to the course sailed. This means that:

- a (small) part of the drag slows the boat.
- the other part of the drag of the sail is involved in the vessel's heel and leeway, o.
- much of the lift of the sail contributes to the advancement of the vessel,
- the other part of the lift of the sail is involved in the vessel's heel and leeway.

Sailing high to the wind generates a perpendicular heeling force. Naval architects plan optimum heel to give maximum forward drive. Technical means used to counter the list include ballast, hydrofoils and counter ballasted keels. Heel can be almost completely offset by the counter- heel technology such as boom / swing keel, type of hydrofoil, etc. These technologies are costly in money, weight, complexity and speed of change of control, so they are reserved for elite competition. In normal cases, the heel remains as extra ballast begins to decrease forward drive. The architect must find a compromise between the amount of resources used to reduce the heel and the heel remaining reasonable. The naval architect often sets the optimal heel between 10° and 20° for monohulls. As a result, the sailor must stick as much as possible to the best heel chosen by the architect. Less heel may mean that the boat is not allowing maximum sail performance. More heel means that the head of the sail drops, thus reducing pressure, in which case the sailing profile is not the best.

The sailor desires optimum heel, a heel giving optimum perpendicular force for the best resulting driving force and to minimise the ratio of perpendicular force to driving force. This ratio depends on the point of sail, incidence, the drag and lift for a given profile. As the lift is the main contributor to the force that drives the boat, and drag usually the main contributor to perpendicular heeling and leeway forces, it is desirable to maximise the L/D. The point of sail depends on the course chosen by the sailor.

The point of sail is a fixed parameter, not a variable of optimisation. But each apparent wind angle relative to the axis of the ship has a different optimum settings. The sail trimmer will first select the trim profile giving maximum lift. Each profile corresponds to a different polar diagram. A sail is generally flexible, the sailor changes the trim through:

- the position of the draft of the sail by adjusting the elements acting on the tension of the fabric of the sail
- adjusting twist of the sail using the leech tension lines such as boom vang or jib sheet angle adjustment.

Many polar curves exist for the possible sail twists and draft positions. The goal is choosing the optimal one. The twist will be set for constant incidence angle along the luff for maximum sail performance, remembering that apparent wind strength and angle varies with altitude. The best L/D is usually obtained when the draft is as far forward as possible. The more forward the draft, the greater the angle of incidence over the luff area.

There comes an attack angle when the air streams do not stick to the sail, creating a dead zone of turbulence which reduces the efficiency of the sail. This inefficient zone is located just after the luff on the windward side. The tell tales in this area become unstable. The flatter the stretched fabric over the sail is, the less the draft. The yacht has several elements acting on the tension of the fabric of the sail:

- Cunningham tension,
- the tack,
- the head point,
- the clew of the sail.
- the backstay,
- shrouds. They act indirectly.

These elements can interact. For example, backstay tension also affects the tension of the head point and therefore the shape of the luff. Both high clew sheet tension tightening the foot and a tighter backstay cause slackening of the leech. For a flexible sail, the camber of the sail and position of the draft are linked. This is a result of their dependence on shape of the cut of the sail.

The camber is a major factor for maximising lift. It is the naval architect or sail maker that sets the cut of the sail for the draft-camber relationship. The thickness of the airfoil profile corresponds to the thickness of sail fabric. Variations in thickness of a sail are negligible compared to the dimensions of the sail. Sail thickness is not a variable to optimise. Contrast mast thickness and profile which are much more important.

For the naval architect the sail-shape offering a large L/D is one with a large aspect ratio. (see previous polar diagram) This explains why modern boats use the Bermudan rig.

Sail drag has three influences:

- induced drag (see influence of aspect ratio on the lift). As the profile is not of infinite length, the ends of the sail, foot and head, equalise the depression of the leeward surface with the pressure of the upwind surface. This dissipated pressure balance becomes the induced drag.
- friction drag, related to boundary layer laminar turbulent flow and roughness of fabric
- form drag, related to choice of the airfoil profile, camber, draft position, and mast profile
- (parasitic drag is related to parts extraneous to the sail, but may influence rigging, boat and sail design)

Prandtl's lift theory applied to thin profile is less complex than the resolution of Navier-Stokes equations, but clearly explains the aspect ratio's effect on induced drag. It shows that the principal factor influencing L/D is induced drag. This theory is very close to reality for a low-impact thin profile. Smaller secondary terms include the form drag and friction drag. This theory shows that the factor with main influence is aspect ratio. The architect chooses the best aspect ratio for best sailing, confirming the choice of Bermudan rig. The sailor's choice of sail trim affects the factors of secondary importance.

POWER. DOWNWIND SAIL CUT AND TRIM

Downwind sailing forces tend to pitch the boat forward. Heeling (rolling axis) forces are less important (in theoretical steady state conditions only). The apparent wind is at an acutely aft angle to the axis of the ship. The chord of the sail is roughly square to the axis of the ship. So:

- much of the sail's drag contributes to the advancement of the ship
- the other part of the sail drag is involved in the vessel's heel
- much of the lift is involved in slowing the vessel,
- the other part of the lift is involved in the vessel's heel.

The optimum setting depends on the apparent wind angle relative to the course. The sail profile is chosen for maximum drag. The heel is not a big factor reducing boat speed. The L/D is not a factor in applying the right profile. The overriding factor is to get the sail profile to give the maximum forward drive based on drag or "power". To maximize the power or maximize propulsive effort are equivalent.

Explanation

The polar "power" plots have a higher maximum propulsive effort compared to polar "L/D" plots. The polar plot giving the maximum drag is a draft located behind the sail. Unlike the optimum setting for close hauled, there is no sudden drop in pressure if the trough is set a little too far. The setting of the sail is wider, more tolerant .

The power of the sail depends almost solely on the part of the sail force contributing to the advancement of the ship (along the axis of vessel speed or course made good). The power is treated as part of the sail force contributing to the advancement of the boat. The power is determined by the polar plot of the sail. The polar plot is independent of the apparent wind speed. Nor, in steady state theory as opposed to dynamic reality, does the heel on the sail intervene with the speed setting. So the heel is not taken into account in the polar plot (same for the L/D of a polar plot). The profile of maximum power is not the profile of maximum L/D, where a setting of "power" creates too much heel, a fairly standard error.

SEVERAL SAILS: MULTIDIMENSIONAL PROBLEM RESOLUTION

The previous method for estimating the thrust of each sail is not valid for boats with multiple sails, but it remains a good approximation. Sails close to each other influence each other. A two-dimensional model explains the phenomenon. In the case of a sloop-rigged sailboat, the foresail changes air flow entering onto the mainsail. The conditions of a stable fluid, constant and uniform, necessary for tables which give lift coefficient, are not respected with multiple sails. The cumulative effect of several sails on a boat can be positive or negative. It is well known that for the same total surface sail, two sails properly set are more effective than a single sail set correctly. Two sails can increase the sailing thrust 20% compared a single sail of same area.

MARINE STEAM ENGINE

A marine steam engine is a steam engine that is used to power a ship or boat. This article deals mainly with marine steam engines of the reciprocating type, which were in use from the inception of the steamboat in the early 19th century to their last years of large-scale manufacture during World War II. Reciprocating steam engines were progressively replaced in marine applications during the 20th century by steam turbines and diesel engines.

HISTORY

The first commercially successful steam engine was developed by Thomas

Newcomen in 1712. The steam engine improvements brought forth by James Watt in the later half of the 18th century greatly improved steam engine efficiency and allowed more compact engine arrangements. Successful adaptation of the steam engine to marine applications in England would have to wait until almost a century later after Newcomen, when Scottish engineer William Symington built the world's "first practical steamboat", the Charlotte Dundas, in 1802.

In 1807, the American Robert Fulton built the world's first commercially successful steamboat, simply known as the North River Steamboat, and powered by a Watt engine. Following Fulton's success, steamboat technology developed rapidly on both sides of the Atlantic. Steamboats initially had a short range and were not particularly seaworthy due to their weight, lack of horsepower, and tendency to break down, but they were employed successfully along rivers and canals, and for short journeys along the coast. The first successful transatlantic crossing by a steamship occurred in 1819 when Savannah sailed from Savannah, Georgia to Liverpool, England. The first steamship to make regular transatlantic crossings was the sidewheel steamer Great Western in 1838.

As the 19th century progressed, marine steam engine and steamship technology developed alongside it. Paddle propulsion gradually gave way to the screw propeller, and the introduction of iron and later steel hulls to replace the traditional wooden hull allowed ships to grow ever larger, necessitating steam power plants that were increasingly complex and powerful.

TYPES OF MARINE STEAM ENGINE

A wide variety of reciprocating marine steam engines were developed over the course of the 19th century. The two main methods of classifying such engines are by connection mechanism and cylinder technology. Most early marine engines had the same cylinder technology (simple expansion, see below) but a number of different methods of supplying power to the crankshaft (i.e. connection mechanism) were in use. Thus, early marine engines are classified mostly according to their connection mechanism. Some common connection mechanisms were side-lever, steeple, walking beam and direct-acting (see following sections).

However, steam engines can also be classified according to their cylinder technology (simple expansion, compound, annular etc.). One can therefore sometimes find examples of engines which were classified under both methods, such as the compound walking beam (compound being the cylinder technology and walking beam being the connection method). Over time, as most engines became direct-acting but cylinder technologies were growing more complex, engines began to be classified solely according to their cylinder technology instead. Some of the more commonly encountered types of marine steam engine are listed in the following sections. Note that not all of these terms may have been used exclusively in relation to marine applications.

ENGINES CLASSIFIED BY CONNECTION MECHANISM

Side-lever

The side-lever engine was the first type of steam engine to be widely adopted for marine use in Europe. In the early years of steam navigation, the side-lever was the most common type of marine engine for inland waterway and coastal service in Europe, and it remained for many years the preferred engine for oceangoing service on both sides of the Atlantic.

The side-lever was an adaptation of the earliest form of steam engine, the beam engine. The typical side-lever engine had a pair of heavy horizontal iron beams, known as side-levers, each secured in the centre by a pin near the base of the engine, allowing the levers to pivot through a limited arc. The engine cylinder stood vertically between this pair of levers at one end, with the piston rod attached to a horizontal crosshead, from each end of which a vertical rod, known as a side-rod, extended down each side of the cylinder to connect to the end of the side-lever on the same side. The far ends of the two side-levers were connected to one another by a horizontal crosstail, from which extended a single, common connecting rod which operated the crankshaft as the levers rocked up and down around the central pin.

The main disadvantages of the side-lever engine were that it was large and heavy, and for inland waterway and coastal service, it was soon replaced by lighter and more efficient designs. It remained the dominant engine type for oceangoing service through much of the first half of the 19th century however, due to its relatively low centre of gravity which gave ships more stability in heavy seas. It was also a common early engine type for warships, since its relatively low height made it less susceptible to battle damage. The side-lever engine was a paddlewheel engine and was not suitable for driving screw propellers. The last ship built for transatlantic service to be fitted with a side-lever engine was the Cunard Line's paddle steamer RMS Scotia, considered an anachronism when it entered service in 1862.

Grasshopper

The grasshopper or 'half-lever' engine was a variant of the side-lever engine. The grasshopper engine differs from the conventional side-lever in that the location of the lever pivot and connecting rod are more or less reversed, with the pivot located at one end of the lever instead of the centre, while the connecting rod is attached to the lever between the cylinder at one end and the pivot at the other. Chief advantages of the grasshopper engine were cheapness of construction and robustness, with the type said to require less maintenance than any other type of marine steam engine. Another advantage is that the engine could be easily started from any crank position. Like the conventional side-lever engine however, grasshopper engines were disadvantaged by their weight and size. They were mainly used in small watercraft such as riverboats and tugs.

Crosshead (square)

The crosshead engine, also known as a square, sawmill or A-frame engine, was a type of paddlewheel engine used in the United States. It was the most common type of engine in the early years of American steam navigation. The crosshead engine is described as having a vertical cylinder above the crankshaft, with the piston rod secured to a horizontal crosshead, from each end of which, on opposite sides of the cylinder, extended a connecting rod which rotated its own separate crankshaft. The crosshead operated within vertical guides that enabled the assembly to maintain the correct path as it moved. The engine's alternative name "A-frame" presumably derived from the shape of the frames supporting these guides. Some crosshead engines had more than one cylinder, in which case the piston rods were usually all connected to the same crosshead. An unusual feature of early examples of this type of engine was the installation of flywheels-geared to the crankshafts-which were thought necessary to ensure smooth operation. These gears could apparently be very noisy in operation.

Because the cylinder was placed above the crankshaft in this type of engine, it had a high center of gravity and was therefore deemed unsuitable for oceangoing service, so that its use was largely confined to vessels plying inland waterways. As marine engines grew steadily larger and heavier through the course of the century, the high center of gravity of square crosshead engines became increasingly impractical, leading to their abandonment by the 1840s in favor of the walking beam engine.

The name of this engine can sometimes lead to confusion as "crosshead" is also an alternative name for the steeple engine (see below). Many sources thus prefer to refer to it by its informal name of "square" engine to avoid confusion. Additionally, the marine crosshead or square engine described in this section should not be confused with the term "square engine" as applied to internal combustion engines, which in the latter case refers to an engine whose bore is equal to its stroke.

Walking beam

The walking beam, also known as a "vertical beam", "overhead beam", or simply "beam", was another early adaptation of the beam engine, but its use was confined almost entirely to the United States. After its introduction, the walking beam quickly became the most popular engine type in America for inland waterway and coastal service, and the type proved to have remarkable longevity, with walking beam engines still being occasionally manufactured as late as the 1940s. In marine applications, the beam itself was generally reinforced with iron struts that gave it a characteristic diamond shape, although the supports on which the beam rested were often built of wood. The adjective "walking" was applied because the beam, which rose high above the ship's deck, could be seen operating, and its rocking motion was (somewhat fancifully) likened to a walking motion.

Walking beam engines were a type of paddlewheel engine and were rarely used for powering propellers. They were used primarily for ships and boats working in rivers, lakes and along the coastline, but were a less popular choice for seagoing vessels because the great height of the engine made the vessel less stable in heavy seas. They were also of limited use militarily, because the engine was exposed to enemy fire and could thus be easily disabled. Their popularity in the United States was due primarily to the fact that the walking beam engine was well suited for the shallow-draft boats which operated in America's shallow coastal and inland waterways.

Walking beam engines remained popular with American shipping lines and excursion operations right into the early 20th century. Although the walking beam engine was technically obsolete in the later 19th century, it remained popular with excursion steamer passengers who expected to see the "walking beam" in motion. There were also technical reasons for retaining the walking beam engine in America, as it was easier to build, requiring less precision in its construction. Wood could be used for the main frame of the engine, at a much lower cost than typical practice of using iron castings for more modern engine designs. Fuel was also much cheaper in America than in Europe, so the lower efficiency of the walking beam engine was less of a consideration. The Philadelphian shipbuilder Charles H. Cramp blamed America's general lack of competitiveness with the British shipbuilding industry in the mid-to-late 19th century upon the conservatism of American domestic shipbuilders and shipping line owners, who doggedly clung to outdated technologies like the walking beam and its associated paddlewheel long after they had been abandoned in other parts of the world.

Steeple

The steeple engine, sometimes referred to as a "crosshead" engine, was an early attempt to break away from the beam concept common to both the walking beam and side-lever types, and come up with a smaller, lighter, more efficient design. In a steeple engine, the vertical oscillation of the piston is not converted to a horizontal rocking motion as in a beam engine, but is instead used to move an assembly, composed of a crosshead and two rods, through a vertical guide at the top of the engine which in turn rotates the crankshaft connecting rod below. In early examples of the type, the crosshead assembly was rectangular in shape, but over time it was refined into an elongated triangle. The triangular assembly above the engine cylinder gives the engine its characteristic "steeple" shape, hence the name. Steeple engines were tall like walking beam engines, but much narrower laterally, saving both space and weight. Because of their height and high centre of gravity, they were, like walking beams, considered to be less appropriate for oceangoing service, but they remained highly popular for several decades, especially in Europe, for inland waterway and coastal vessels.

Steeple engines began to appear in steamships in the 1830s and the type was perfected in the early 1840s by the British shipbuilder David Napier. The steeple engine was gradually superseded by the various types of direct-acting engine.

Siamese

The Siamese engine, also referred to as the "double cylinder" or "twin cylinder" engine, was another early alternative to the beam or side-lever engine. This type of engine had two identical, vertical engine cylinders arranged side-by-side, whose piston rods were attached to a common, T-shaped crosshead.

The vertical arm of the crosshead extended down between the two cylinders and was attached at the bottom to both the crankshaft connecting rod and to a guide block that slid between the vertical sides of the cylinders, enabling the assembly to maintain the correct path as it moved. The Siamese engine was invented by British engineer Joseph Maudslay (son of Henry), but although he invented it after his oscillating engine (see below), it failed to achieve the same widespread acceptance, as it was only marginally smaller and lighter than the side-lever engines it was designed to replace. It was however used on a number of mid-century warships, including the first warship fitted with a screw propeller, HMS Rattler.

Direct acting

There are two definitions of a direct-acting engine encountered in 19th-century literature. The earlier definition applies the term "direct-acting" to any type of engine other than a beam (i.e. walking beam, side-lever or grasshopper) engine. The later definition only uses the term for engines which apply their power directly to the crankshaft via the piston rod and/or connecting rod. Unless otherwise noted, this article uses the later definition. Unlike the side-lever or beam engine, a direct-acting engine could be readily adapted to power either paddlewheels or a propeller. As well as offering a lower profile, direct-acting engines had the advantage of being smaller and weighing considerably less than beam or side-lever engines. The Royal Navy found that on average a direct-acting engine (early definition) weighed 40% less and required an engine room only two thirds the size of that for a side-lever of equivalent power. One disadvantage of such engines is that they were more prone to wear and tear and thus required more maintenance.

Oscillating

An oscillating engine was a type of direct-acting engine that was designed to achieve further reductions in engine size and weight. Oscillating engines had the piston rods connected directly to the crankshaft, dispensing with the need for connecting rods. In order to achieve this aim, the engine cylinders were not immobile as in most engines, but secured in the middle by trunnions which allowed the cylinders themselves to pivot back and forth as the crankshaft rotated, hence the term oscillating. Steam was supplied and exhausted through the trunnions. The oscillating motion of the cylinder was usually used to line up ports in the trunnions to direct the steam feed and exhaust to the cylinder at the correct times. However, separate valves may be provided, controlled by the oscillating motion. This allows the timing to be varied to enable expansive working. (As, for example, the engine in the paddle ship PD Krippen.) This compromises the advantage of simplicity but still retains the advantage of compactness. The first patented oscillating engine was built by Joseph Maudslay in 1827, but the type is considered to have been perfected by John Penn. Oscillating engines remained a popular type of marine engine for much of the 19th century.

Trunk

The trunk engine, another type of direct-acting engine, was originally developed as a means of reducing an engine's height while retaining a long stroke (a long stroke was considered important at this time because it reduced the strain on components). Early examples of trunk engines had vertical cylinders; however, it was quickly realized that the type was compact enough to be laid horizontally across the keel. In this configuration, it was very useful to navies, as it had a profile low enough to fit entirely below a ship's waterline, where it would be as safe as possible from enemy fire. The type was generally produced for military service by John Penn.

This type of engine had a horizontal cylinder, through the centre of which passed a cylindrical "trunk" or passage containing the connecting rod. The walls of the trunk were either bolted to the piston or cast as one piece with it, and moved back and forth with it. The working portion of the cylinder was annular or ring-shaped, with the trunk passing through the centre of the cylinder itself.

Trunk engines were quite common on mid-19th century warships, and were also to be found in commercial vessels, where though valued for their compact size and low centre of gravity, they proved expensive to operate. Trunk engines however proved poorly adapted to the higher boiler pressures that became prevalent in the latter half of the 19th century, and were abandoned in favour of other solutions. Normally large engines, a small mass-produced, high-revolution, high-pressure version was produced for the Crimean War. In being quite effective, the type persisted in later gunboats.

An original trunk engine of the gunboat type exists in the Western Australian Museum in Fremantle. After sinking in 1872, it was raised in 1985 from the SS Xantho and can now be turned over by hand. The engine's mode of operation, illustrating its compact nature, can be viewed on the Xantho project's website.

Vibrating lever

The vibrating lever, or half-trunk engine, was a development of the conventional trunk engine conceived by Swedish-American engineer John Ericsson. Ericsson needed a small, low-profile engine like the trunk engine to power the U.S. Federal government's monitors, a type of warship developed during the American Civil War that had very little space for a conventional powerplant. The trunk engine itself was however unsuitable for this purpose because the preponderance of weight was on the side of the engine containing the cylinder and trunk, a problem which could not be compensated for on the small monitor warships.

Ericsson resolved this problem by placing two horizontal cylinders back-to-back in the middle of the engine, working two "vibrating levers", one on each side, which by means of shafts and additional levers rotated a centrally located crankshaft. Vibrating lever engines were later used in some other warships and merchant vessels, but their use was confined to ships built in the United States and in Ericsson's native country of Sweden, and as they had few advantages over more conventional engines, were soon supplanted by other types.

Back acting

The back-acting engine, also known as the return connecting rod engine, was another engine designed to have a very low profile. The back-acting engine was in effect a modified steeple engine, laid horizontally across the keel of a ship rather than standing vertically above it.

Instead of the triangular crosshead assembly found in a typical steeple engine however, the back-acting engine generally utilized a set of two or more elongated, parallel piston rods terminating in a crosshead to perform the same function. The term "back-acting" or "return connecting rod" derives from the fact that the connecting rod "returns" or comes back from the side of the engine opposite the engine cylinder to rotate a centrally located crankshaft. Back-acting engines were another type of engine popular in both warships and commercial vessels in the mid-19th century, but like many other engine types in this era of rapidly changing technology, they were eventually abandoned for other solutions.

There is only one back-acting engine known to be still in existence-that of the TV Emery Rice (formerly USS Ranger), now the centerpiece of a display at the American Merchant Marine Museum.

Vertical

As steamships grew steadily in size and tonnage through the course of the 19th century, the need for low profile, low centre-of-gravity engines correspondingly declined. Freed increasingly from these design constraints, engineers were able to revert to simpler, more efficient and more easily maintained designs. The result was the growing dominance of the so-called "vertical" engine (more correctly known as the vertical inverted direct acting engine).

In this type of engine, the cylinders are located directly above the crankshaft, with the piston rod/connecting rod assemblies forming a more or less straight line between the two. The configuration is similar to that of a modern internal combustion engine (one notable difference being that the steam engine is double acting, see below, whereas an internal combustion engine generates power only in the downward stroke). Vertical engines are sometimes referred to as "hammer", "forge hammer" or "steam hammer" engines, due to their roughly similar appearance to another common 19th-century steam technology, the steam hammer.

Vertical engines came to supersede almost every other type of marine steam engine toward the close of the 19th century. Because they became so common, vertical engines are not usually referred to as such, but are instead referred to based upon their cylinder technology, i.e. as compound, triple expansion, quadruple expansion etc. It should be noted that the term "vertical" for this type of engine is imprecise, since technically any type of steam engine is "vertical" if the cylinder is vertically oriented. An engine described as "vertical" should therefore not be assumed to be of the vertical inverted direct-acting type unless the term "vertical" is unqualified.

ENGINES CLASSIFIED BY CYLINDER TECHNOLOGY

Simple expansion

A simple expansion engine is a steam engine that expands the steam through only one stage, which is to say, all its cylinders are operated at the same pressure. Since this was by far the most common type of engine in the early period of marine engine development, the term "simple expansion" is rarely encountered; rather, an engine is assumed to be simple expansion unless otherwise stated.

Compound

A compound engine is a steam engine which operates cylinders through more than one stage, i.e., at different pressure levels. Compound engines were a method of improving efficiency. Up until the development of compound engines, steam engines used the steam only once before being recycled back to the boiler, but a compound engine recycles the steam into one or more larger, lower pressure second cylinders first, in order to utilize more of its heat energy. Compound engines could be configured to either increase a ship's economy or its speed. Although broadly speaking a compound engine can refer to a steam engine with any number of different-pressure cylinders, the term usually refers to engines which expand steam through only two stages, i.e. those which operate cylinders at only two different pressures (or "double expansion" engines).

Note that a compound engine (including multiple expansion engines, see below) can have more than one set of variable-pressure cylinders. For example, an engine might have two cylinders operating at pressure x and two operating at pressure y, or one cylinder operating at pressure x and three operating at pressure y. What makes it compound (or double expansion) as opposed to multiple expansion is that there are only two pressures, x and y.

The first compound engine believed to have been installed in a ship was that fitted to Henry Eckford by the American engineer James P. Allaire in 1824. However, many sources attribute the "invention" of the marine compound engine to Glasgow's John Elder in the 1850s. Elder made improvements to the compound engine that made it safe and economical for ocean-crossing voyages for the first time.

Triple or multiple expansion

A triple expansion engine is a compound engine that expands the steam in three stages, i.e. an engine which has cylinders operating at three different pressures. A quadruple expansion engine expands the steam in four stages, and so on. The first successful commercial use was an engine built at Govan in Scotland by Alexander C. Kirk for the SS Aberdeen in 1881.

Multiple expansion engine manufacture continued well into the 20th century. All 2,700 Liberty ships built by the United States during World War II were powered by triple-expansion engines, because the capacity of the US to manufacture steam turbines and diesels was still limited. The biggest manufacturer of triple expansion engines during the war was the Joshua Hendy Iron Works. Toward the end of the war, turbine-powered Victory ships were manufactured in increasing numbers.

Annular

An annular engine is an unusual type of engine that has an annular (ring-shaped) cylinder. Some of American pioneering engineer James P. Allaire's early compound engines were of the annular type, with a smaller, high pressure cylinder placed in the centre of a larger, ring-shaped low-pressure cylinder.

Trunk engines were another type of annular engine. A third type of annular marine engine which was sometimes produced utilized the Siamese engine connecting mechanism, but instead of two separate cylinders, had a single annular-shaped cylinder wrapped around the vertical arm of the crosshead (see diagram under "Siamese" above).

OTHER TERMS

Some other terms are encountered in marine engine literature of the period. These terms, listed below, are usually used in conjunction with one or more of the basic engine classification terms listed above.

Simple

A simple engine is an engine with only one cylinder. Up until about the mid-19th century, most ships had engines with only one cylinder (although some vessels had more than one engine). Simple engines are always also simple expansion engines by necessity.

Double acting

A double acting engine is an engine where steam is applied to both the up and down stroke of the piston. Earlier steam engines applied steam in only one direction, allowing momentum or gravity to return the piston to its starting place, but a double acting engine uses steam to force the piston in both directions, thus increasing RPM and power.

Like the term "simple expansion", the term "double acting" is infrequently encountered in the literature since almost all marine engines were of the double acting type.

Vertical, horizontal, inclined, inverted

These terms refer to the orientation of the engine cylinder. A vertical cylinder stands vertically with its piston rod operating above it. An inverted cylinder (or "vertical inverted" cylinder) can be thought of as a vertical cylinder positioned upside down.

With an inclined or horizontal type, the cylinder and piston are positioned at an incline or horizontally. An inclined inverted cylinder is an inverted cylinder operating at an incline.

These terms are all generally used in conjunction with the engine types above. Thus, one may have a horizontal direct-acting engine, or an inverted walking beam, and so on. Inclined and horizontal cylinders could be very useful in naval vessels as their orientation kept the engine profile as low as possible and thus less susceptible to damage. They could also be used in a low profile ship or to keep a ship's centre of gravity lower. In addition, inclined or horizontal cylinders had the advantage of reducing the amount of vibration by comparison with a vertical cylinder.

Geared

A geared engine or "geared screw" turns the propeller at a different rate to the engine's RPM. Early marine propeller engines were geared upward, which is to say the propeller was geared to run at a higher RPM than the engine itself. As engines became faster and more powerful through the latter part of the 19th century, gearing was often dispensed with and the propeller ran at the same RPM as the engine.

3

Underwater

Underwater is a term describing the realm below the surface of water where the water exists in a natural feature (called a body of water) such as an ocean, sea, lake, pond, or river. Three quarters of the planet Earth is covered by water. A majority of the planet's solid surface is abyssal plain, at depths between 4,000 and 5,500 metres (13,000 and 18,000 ft) below the surface of the oceans.

The solid surface location on the planet closest to the centre of the orb is the Challenger Deep, located in the Mariana Trench at a depth of 10,924 metres (35,840 ft). Although a number of human activities are conducted underwater-such as research, scuba diving for work or recreation, or even underwater warfare with submarines, this very extensive environment on planet Earth is hostile to humans in many ways and therefore little explored. But it can be explored by sonar, or more directly via manned or autonomous submersibles. The ocean floors have been surveyed via sonar to at least a coarse resolution; particularly-strategic areas have been mapped in detail, in the name of detecting enemy submarines, or aiding friendly ones, though the resulting maps may still be classified.

An immediate obstacle to human activity under water is the fact that human lungs cannot naturally function in this environment. Unlike the gills of fish, human lungs are adapted to the exchange of gases at atmospheric pressure, not liquids. Aside from simply having insufficient musculature to rapidly move water in and out of the lungs, a more significant problem for all air-breathing animals, such as mammals and birds, is that water contains so little dissolved oxygen compared with atmospheric air. Air is around 21% O_2; water typically is less than 0.001% dissolved oxygen.

The density of water also causes problems that increase dramatically with depth. The atmospheric pressure at the surface is 14.7 pounds per square inch or around 100 kPa. A comparable water pressure occurs at a depth of only 10 m (33 ft) (9.8 m (32 ft) for sea water). Thus, at about 10 m below the surface, the water exerts twice the pressure (2 atmospheres or 200 kPa) on the body as air at surface level. For solid objects like human bones and muscles, this added pressure is not much of a problem; but it is a problem for any air-filled spaces like the mouth, ears, paranasal sinuses and lungs. This is because the air in those spaces reduces in volume when under pressure and so does not provide those spaces with support from the higher outside pressure. Even at a depth of 8 ft (2.4 m) underwater, an inability to equalize air pressure in the middle ear with outside water pressure can cause pain, and the tympanic membrane can rupture at depths under 10 ft (3 m). The danger of pressure damage is greatest in shallow water because the rate of pressure change is greatest at the surface of the water. For example the pressure increase between the surface and 10 m (33 ft) is 100% (100 kPa to 200 kPa), but the pressure increase from 30 m (100 ft) to 40 m (130 ft) is only 25% (400 kPa to 500 kPa).

Any object immersed in water is provided with a buoyant force that counters the force of gravity, appearing to make the object less heavy. If the overall density of the object exceeds the density of water, the object sinks. If the overall density is less than the density of water, the object rises until it floats on the surface.

With increasing depth underwater, sunlight is absorbed, and the amount of visible light diminishes. Because absorption is greater for long wavelengths (red end of the visible spectrum) than for short wavelengths (blue end of the visible spectrum), the colour spectrum is rapidly altered with increasing depth. White objects at the surface appear bluish underwater, and red objects appear dark, even black. Although light penetration will be less if water is turbid, in the very clear water of the open ocean less than 25% of the surface light reaches a depth of 10 m (33 feet). At 100 m (330 ft) the light present from the sun is normally about 0.5% of that at the surface.

The euphotic depth is the depth at which light intensity falls to 1% of the value at the surface. This depth is dependent upon water clarity, being only a few metres underwater in a turbid estuary, but may reach up to 200 metres in the open ocean. At the euphotic depth, plants (such as phytoplankton) have no net energy gain from photosynthesis and thus cannot grow. At depths greater than a few hundred metres, the sun has little effect on water temperature, because the sun's energy has been absorbed by water at the surface. In the great depths of the ocean the water temperature is very low. In fact, 75% of the water in the world ocean (the great depths) has a temperature between 0 °C and 2 °C.

Water conducts heat around twenty five times more efficiently than air. Hypothermia, a potentially fatal condition, occurs when the human body's core temperature falls below 35 °C. Insulating the body's warmth from water is the main purpose of diving suits and exposure suits when used in water temperatures below 25 °C.

Sound is transmitted about 4.3 times faster in water (1,484 m/s in fresh water) as it is in air (343 m/s). The human brain can determine the direction of sound in air by detecting small differences in the time it takes for sound waves in air to reach each of the two ears. For these reasons divers find it difficult to determine the direction of sound underwater. However, some animals have adapted to this difference and many use sound to navigate underwater.

UNDERWATER ACOUSTICS

Underwater acoustics is the study of the propagation of sound in water and the interaction of the mechanical waves that constitute sound with the water and its boundaries. The water may be in the ocean, a lake or a tank. Typical frequencies associated with underwater acoustics are between 10 Hz and 1 MHz.

The propagation of sound in the ocean at frequencies lower than 10 Hz is usually not possible without penetrating deep into the seabed, whereas frequencies above 1 MHz are rarely used because they are absorbed very quickly. Underwater acoustics is sometimes known as hydroacoustics. The field of underwater acoustics is closely related to a number of other fields of acoustic study, including sonar, transduction, acoustic signal processing, acoustical oceanography, bioacoustics, and physical acoustics.

HISTORY

Underwater sound has probably been used by marine animals for millions of years. The science of underwater acoustics began in 1490, when Leonardo da Vinci wrote, "If you cause your ship to stop and place the head of a long tube in the water and place the outer extremity to your ear, you will hear ships at a great distance from you."

In 1687 Isaac Newton wrote his Mathematical Principles of Natural Philosophy which included the first mathematical treatment of sound. The next major step in the development of underwater acoustics was made by Daniel Colladon, a Swiss physicist, and Charles Sturm, a French mathematician. In 1826, on Lake Geneva, they measured the elapsed time between a flash of light and the sound of a submerged ship's bell heard using an underwater listening horn. They measured a sound speed of 1435 metres per second over a 17 kilometre distance, providing the first quantitative measurement of sound speed in water. The result they obtained was within about 2% of currently accepted values. In 1877 Lord Rayleigh wrote the Theory of Sound and established modern acoustic theory.

The sinking of Titanic in 1912 and the start of World War I provided the impetus for the next wave of progress in underwater acoustics. Systems for detecting icebergs and U-boats were developed. Between 1912 and 1914, a number of echolocation patents were granted in Europe and the U.S., culminating in Reginald A. Fessenden's echo-ranger in 1914. Pioneering work was carried out during this time in France by Paul Langevin and in Britain by A B Wood and associates. The development of both active ASDIC and passive sonar (SOund Navigation And Ranging) proceeded apace during the war, driven by the first large scale deployments of submarines. Other advances in underwater acoustics included the development of acoustic mines. In 1919, the first scientific paper on underwater acoustics was published, theoretically describing the refraction of sound waves produced by temperature and salinity gradients in the ocean. The range predictions of the paper were experimentally validated by transmission loss measurements.

The next two decades saw the development of several applications of underwater acoustics. The fathometer, or depth sounder, was developed commercially during the 1920s. Originally natural materials were used for the transducers, but by the 1930s sonar systems incorporating piezoelectric transducers made from synthetic materials were being used for passive listening systems and for active echo-ranging systems. These systems were used to good effect during World War II by both submarines and anti-submarine vessels. Many advances in underwater acoustics were made which were summarised later in the series Physics of Sound in the Sea, published in 1946. After World War II, the development of sonar systems was driven largely by the Cold War, resulting in advances in the theoretical and practical understanding of underwater acoustics, aided by computer-based techniques.

THEORY

SOUND WAVES IN WATER

A sound wave propagating underwater consists of alternating compressions and rarefactions of the water. These compressions and rarefactions are detected by a receiver, such as the human ear or a hydrophone, as changes in pressure. These waves may be man-made or naturally generated.

SPEED OF SOUND, DENSITY AND IMPEDANCE

The speed of sound c (i.e., the longitudinal motion of wavefronts) is related to frequency f and wavelength o#of a wave by $c = f.o$. This is different from the particle velocity u, which refers to the motion of molecules in the medium due to the sound, and relates the plane wave the pressure p to the fluid density u#and sound speed c by $p = c\ .\ u\ .$ u. The product of c and u#from the above formula is known as the characteristic acoustic impedance.

The acoustic power (energy per second) crossing unit area is known as the intensity of the wave and for a plane wave the average intensity is given by $I \ @\ q^2 / (uc)$, where q is the root mean square acoustic pressure. At 1 kHz, the wavelength in water is about 1.5 m. Sometimes the term "sound velocity" is used but this is incorrect as the quantity is a scalar. The large impedance contrast between air and water (the ratio is about 3600) and the scale of surface roughness means that the sea surface behaves as an almost perfect reflector of sound at frequencies below 1 kHz. Sound speed in water exceeds that in air by a factor of 4.4 and the density ratio is about 820.

ABSORPTION OF SOUND

Absorption of low frequency sound is weak. (see Technical Guides - Calculation of absorption of sound in seawater for an on-line calculator). The main cause of sound attenuation in fresh water, and at high frequency in sea water (above 100 kHz) is viscosity. Important additional contributions at lower frequency in seawater are associated with the ionic relaxation of boric acid (up to c. 10 kHz) and magnesium sulfate (c. 10 kHz-500 kHz). Sound may be absorbed by losses at the fluid boundaries. Near the surface of the sea losses can occur in a bubble layer or in ice, while at the bottom sound can penetrate into the sediment and be absorbed.

SOUND REFLECTION AND SCATTERING

Boundary interactions

Both the water surface and bottom are reflecting and scattering boundaries.

Surface

For many purposes the sea-air surface can be thought of as a perfect reflector. The impedance contrast is so great that little energy is able to cross this boundary. Acoustic pressure waves reflected from the sea surface experience a reversal in phase, often stated as either a "pi phase change" or a "180 deg phase change". This is represented mathematically by assigning a reflection coefficient of minus 1 instead of plus one to the sea surface.

At high frequency (above about 1 kHz) or when the sea is rough, some of the incident sound is scattered, and this is taken into account by assigning a reflection coefficient whose magnitude is less than one. For example, close to normal incidence, the reflection coefficient becomes $R @ 0e^{02k^2h^2 \sin^2 A}$, where h is the rms wave height. A further complication is the presence of wind generated bubbles or fish close to the sea surface. The bubbles can also form plumes that absorb some of the incident and scattered sound, and scatter some of the sound themselves.

Seabed

The acoustic impedance mismatch between water and the bottom is generally much less than at the surface and is more complex. It depends on the bottom material types and depth of the layers. Theories have been developed for predicting the sound propagation in the bottom in this case, for example by Biot and by Buckingham.

At Target

The reflection of sound at a target whose dimensions are large compared with the acoustic wavelength depends on its size and shape as well as the impedance of the target relative to that of water. Formulae have been developed for the target strength of various simple shapes as a function of angle of sound incidence. More complex shapes may be approximated by combining these simple ones.

PROPAGATION OF SOUND

Underwater acoustic propagation depends on many factors. The direction of sound propagation is determined by the sound speed gradients in the water. In the sea the vertical gradients are generally much larger than the horizontal ones. Combining this with a tendency towards increasing sound speed at increasing depth, due to the increasing pressure in the deep sea, causes a reversal of the sound speed gradient in the thermocline, creating an efficient waveguide at the depth, corresponding to the minimum sound speed. The sound speed profile may cause regions of low sound intensity called "Shadow Zones," and regions of high intensity called "Caustics". These may be found by ray tracing methods.

At equator and temperate latitudes in the ocean, the surface temperature is high enough to reverse the pressure effect, such that a sound speed minimum occurs at depth of a few hundred metres. The presence of this minimum creates a special channel known as Deep Sound Channel, previously known as the SOFAR (sound fixing and ranging) channel, permitting guided propagation of underwater sound for thousands of kilometres without interaction with the sea surface or the seabed. Another phenomenon in the deep sea is the formation of sound focusing areas, known as Convergence Zones. In this case sound is refracted downward from a near-surface source and then back up again. The horizontal distance from the source at which this occurs depends on the positive and negative sound speed gradients. A surface duct can also occur in both deep and moderately shallow water when there is upward refraction, for example due to cold surface temperatures. Propagation is by repeated sound bounces off the surface.

In general, as sound propagates underwater there is a reduction in the sound intensity over increasing ranges, though in some circumstances a gain can be obtained due to focusing. Propagation loss (sometimes referred to as transmission loss) is a quantitative measure of the reduction in sound intensity between two points, normally the sound source and a distant receiver. If I_s is the far field intensity of the source referred to a point 1 m from its acoustic centre and I_r is the intensity at the receiver, then the propagation loss is given by PL = $10 log(I_s/I_r)$. In this equation I_r is not the true acoustic intensity at the receiver, which is a vector quantity, but a scalar equal to the equivalent plane wave intensity (EPWI) of the sound field. The EPWI is defined as the magnitude of the intensity of a plane wave of the same RMS pressure as the true acoustic field. At short range the propagation loss is dominated by spreading while at long range it is dominated by absorption and/or scattering losses. An alternative definition is possible in terms of pressure instead of intensity, giving $PL = 20log(p_s/p_r)$, where P_s is the RMS acoustic pressure in the far-field of the projector, scaled to a standard distance of 1 m, and P_r is the RMS pressure at the receiver position.

These two definitions are not exactly equivalent because the characteristic impedance at the receiver may be different from that at the source. Because of this, the use of the intensity definition leads to a different sonar equation to the definition based on a pressure ratio. If the source and receiver are both in water, the difference is small.

PROPAGATION MODELLING

The propagation of sound through water is described by the wave equation, with appropriate boundary conditions. A number of models have been developed to simplify propagation calculations. These models include ray theory, normal mode solutions, and parabolic equation simplifications of the wave equation. Each set of solutions is generally valid and computationally efficient in a limited frequency and range regime, and may involve other limits as well. Ray theory is more appropriate at short range and high frequency, while the other solutions function better at long range and low frequency. Various empirical and analytical formulae have also been derived from measurements that are useful approximations.

REVERBERATION

Transient sounds result in a decaying background that can be of much larger duration than the original transient signal. The cause of this background, known as reverberation, is partly due to scattering from rough boundaries and partly due to scattering from fish and other biota. For an acoustic signal to be detected easily, it must exceed the reverberation level as well as the background noise level.

DOPPLER SHIFT

If an underwater object is moving relative to an underwater receiver, the frequency of the received sound is different from that of the sound radiated (or reflected) by the object. This change in frequency is known as a Doppler shift. The shift can be observed in active sonar systems, particularly narrow-band ones, because the transmitter frequency is known, and the relative motion between sonar and object can be calculated. Sometimes the frequency of the radiated noise (a tonal) may also be known, in which case the same calculation can be done for passive sonar. For active systems the change in frequency is 0.69 Hz per knot per kHz and half this for passive systems as propagation is only one way. The shift corresponds to an increase in frequency for an approaching target.

INTENSITY FLUCTUATIONS

Though acoustic propagation modelling generally predicts a constant received sound level, in practice there are both temporal and spatial fluctuations. These may be due to both small and large scale environmental phenomena. These can include sound speed profile fine structure and frontal zones as well as internal waves. Because in general there are multiple propagation paths between a source and receiver, small phase changes in the interference pattern between these paths can lead to large fluctuations in sound intensity.

NON-LINEARITY

In water, especially with air bubbles, the change in density due to a change in pressure is not exactly linearly proportional. As a consequence for a sinusoidal wave input additional harmonic and subharmonic frequencies are generated. When two sinusoidal waves are input, sum and difference frequencies are generated. The conversion process is greater at high source levels than small ones. Because of the non-linearity there is a dependence of sound speed on the pressure amplitude so that large changes travel faster than small ones. Thus a sinusoidal waveform gradually becomes a sawtooth one with a steep rise and a gradual tail. Use is made of this phenomenon in parametric sonar and theories have been developed to account for this, e.g. by Westerfield.

MEASUREMENTS

Sound in water is measured using a hydrophone, which is the underwater equivalent of a microphone. A hydrophone measures pressure fluctuations, and these are usually converted to sound pressure level (SPL), which is a logarithmic measure of the mean square acoustic pressure.

Measurements are usually reported in one of three forms:

- RMS acoustic pressure in micropascals (or dB re 1 pPa)
- RMS acoustic pressure in a specified bandwidth, usually octaves or thirds of octave (dB re 1 pPa)
- spectral density (mean square pressure per unit bandwidth) in micropascals per hertz (dB re 1 pPa^2/Hz)

SOUND SPEED

Approximate values for fresh water and seawater, respectively, at atmospheric pressure are 1450 and 1500 m/s for the sound speed, and 1000 and 1030 kg/m^3 for the density. The speed of sound in water increases with increasing pressure, temperature and salinity. The maximum speed in pure water under atmospheric pressure is attained at about 74°C; sound travels slower in hotter water after that point; the maximum increases with pressure. On-line calculators can be found at Technical Guides - Speed of Sound in Sea-Water and Technical Guides - Speed of Sound in Pure Water.

ABSORPTION

Many measurements have been made of sound absorption in lakes and the ocean (see Technical Guides - Calculation of absorption of sound in seawater for an on-line calculator).

AMBIENT NOISE

Measurement of acoustic signals are possible if their amplitude exceeds a minimum threshold, determined partly by the signal processing used and partly by the level of background noise. Ambient noise is that part of the received noise that is independent of the source, receiver and platform characteristics.

This it excludes reverberation and towing noise for example. The background noise present in the ocean, or ambient noise, has many different sources and varies with location and frequency. At the lowest frequencies, from about 0.1 Hz to 10 Hz, ocean turbulence and microseisms are the primary contributors to the noise background. Typical noise spectrum levels decrease with increasing frequency from about 140 dB re 1 pPa^2/Hz at 1 Hz to about 30 dB re 1 pPa^2/Hz at 100 kHz. Distant ship traffic is one of the dominant noise sources in most areas for frequencies of around 100 Hz, while wind-induced surface noise is the main source between 1 kHz and 30 kHz. At very high frequencies, above 100 kHz, thermal noise of water molecules begins to dominate. The thermal noise spectral level at 100 kHz is 25 dB re 1 pPa^2/Hz. The spectral density of thermal noise increases by 20 dB per decade (approximately 6 dB per octave).

Transient sound sources also contribute to ambient noise. These can include intermittent geological activity, such as earthquakes and underwater volcanoes, rainfall on the surface, and biological activity. Biological sources include cetaceans (especially blue, fin and sperm whales), certain types of fish, and snapping shrimp. Rain can produce high levels of ambient noise. However the numerical relationship between rain rate and ambient noise level is difficult to determine because measurement of rain rate is problematic at sea.

REVERBERATION

Many measurements have been made of sea surface, bottom and volume reverberation. Empirical models have sometimes been derived from these. A commonly used expression for the band 0.4 to 6.4 kHz is that by Chapman and Harris. It is found that a sinusoidal waveform is spread in frequency due to the surface motion.

For bottom reverberation a Lambert's Law is found often to apply approximately, for example see Mackenzie. Volume reverberation is usually found to occur mainly in layers, which change depth with the time of day, e.g., see Marshall and Chapman. The under-surface of ice can produce strong reverberation when it is rough, see for example Milne.

BOTTOM LOSS

Bottom loss has been measured as a function of grazing angle for many frequencies in various locations, for example those by the US Marine Geophysical Survey. The loss depends on the sound speed in the bottom (which is affected by gradients and layering) and by roughness. Graphs have been produced for the loss to be expected in particular circumstances. In shallow water bottom loss often has the dominant impact on long range propagation. At low frequencies sound can propagate through the sediment then back into the water.

UNDERWATER HEARING

Comparison with airborne sound levels

As with airborne sound, sound pressure level underwater is usually reported in units of decibels, but there are some important differences that make it difficult (and often inappropriate) to compare SPL in water with SPL in air. These differences include:

- difference in reference pressure: 1 pPa (one micropascal, or one millionth of a pascal) instead of 20 pPa.
- difference in interpretation: there are two schools of thought, one maintaining that pressures should be compared directly, and that the other that one should first convert to the intensity of an equivalent plane wave.
- difference in hearing sensitivity: any comparison with (A-weighted) sound in air needs to take into account the differences in hearing sensitivity, either of a human diver or other animal.

Hearing sensitivity

The lowest audible SPL for a human diver with normal hearing is about 67 dB re 1 pPa, with greatest sensitivity occurring at frequencies around 1 kHz. Dolphins and other toothed whales are renowned for their acute hearing sensitivity, especially in the frequency range 5 to 50 kHz. Several species have hearing thresholds between 30 and 50 dB re 1 pPa in this frequency range. For example the hearing threshold of the killer whale occurs at an RMS acoustic pressure of 0.02 mPa (and frequency 15 kHz), corresponding to an SPL threshold of 26 dB re 1 pPa. By comparison the most sensitive fish is the soldier fish, whose threshold is 0.32 mPa (50 dB re 1 pPa) at 1.3 kHz, whereas the lobster has a hearing threshold of 1.3 Pa at 70 Hz (122 dB re 1 pPa).

Safety thresholds

High levels of underwater sound create a potential hazard to marine and amphibious animals as well as to human divers. Guidelines for exposure of human divers and marine mammals to underwater sound are reported by the SOLMAR project of the NATO Undersea Research Centre. Human divers exposed to SPL above 154 dB re 1 ?Pa in the frequency range 0.6 to 2.5 kHz are reported to experience changes in their heart rate or breathing frequency. Diver aversion to low frequency sound is dependent upon sound pressure level and center frequency.

APPLICATIONS OF UNDERWATER ACOUSTICS

Sonar

Sonar is the name given to the acoustic equivalent of radar. Pulses of sound are used to probe the sea, and the echoes are then processed to extract information about the sea, its boundaries and submerged objects. An alternative use, known as passive sonar, attempts to do the same by listening to the sounds radiated by underwater objects.

Underwater communication

The need for underwater acoustic telemetry exists in applications such as data harvesting for environmental monitoring, communication with and between manned and unmanned underwater vehicles, transmission of diver speech, etc. A related application is underwater remote control, in which acoustic telemetry is used to remotely actuate a switch or trigger an event. A prominent example of underwater remote control are acoustic releases, devices that are used to return sea floor deployed instrument packages or other payloads to the surface per remote command at the end of a deployment. Acoustic communications form an active field of research with significant challenges to overcome, especially in horizontal, shallow-water channels. Compared with radio telecommunications, the available bandwidth is reduced by several orders of magnitude. Moreover, the low speed of sound causes multipath propagation to stretch over time delay intervals of tens or hundreds of milliseconds, as well as significant Doppler shifts and spreading. Often acoustic communication systems are not limited by noise, but by reverberation and time variability beyond the capability of receiver algorithms. The fidelity of underwater communication links can be greatly improved by the use of hydrophone arrays, which allow processing techniques such as adaptive beamforming and diversity combining.

Underwater Navigation and Tracking

Underwater navigation and tracking is a common requirement for exploration and work by divers, ROV, autonomous underwater vehicles (AUV), manned submersibles and submarines alike. Unlike most radio signals which are quickly absorbed, sound propagates far underwater and at a rate that can be precisely measured or estimated. It can thus be used to measure distances between a tracked target and one or multiple reference of baseline stations precisely, and triangulate the position of the target, sometimes with centimeter accuracy. Starting in the 1960s, this has given rise to underwater acoustic positioning systems which are now widely used.

Seismic exploration

Seismic exploration involves the use of low frequency sound (< 100 Hz) to probe deep into the seabed. Despite the relatively poor resolution due to their long wavelength, low frequency sounds are preferred because high frequencies are heavily attenuated when they travel through the seabed. Sound sources used include airguns, vibroseis and explosives.

Weather and climate observation

Acoustic sensors can be used to monitor the sound made by wind and precipitation. For example, an acoustic rain gauge is described by Nystuen. Lightning strikes can also be detected.Acoustic thermometry of ocean climate (ATOC) uses low frequency sound to measure the global ocean temperature.

Oceanography

Large scale ocean features can be detected by acoustic tomography. Bottom characteristics can be measured by side-scan sonar and sub-bottom profiling.

Marine biology

Due to its excellent propagation properties, underwater sound is used as a tool to aid the study of marine life, from microplankton to the blue whale. Echo sounders are often used to provide data on marine life abundance, distribution, and behavior information. Echo sounders, also referred to as hydroacoustics is also used for fish location, quantity, size, and biomass.

Acoustic telemetry is also used for monitoring fishes and marine wildlife. An acoustic transmitter is attached to the fish (sometimes internally) while an array of receivers listen to the information conveyed by the sound wave. This enables the researchers to track the movements of individuals in a small-medium scale.

Particle physics

A neutrino is a fundamental particle that interacts very weakly with other matter. For this reason, it requires detection apparatus on a very large scale, and the ocean is sometimes used for this purpose. In particular, it is thought that ultra-high energy neutrinos in seawater can be detected acoustically.

4

Offshore Construction

Offshore construction is the installation of structures and facilities in a marine environment, usually for the production and transmission of electricity, oil, gas and other resources. Construction and pre-commissioning is typically performed as much as possible onshore. To optimize the costs and risks of installing large offshore platforms, different construction strategies have been developed.

One strategy is to fully construct the offshore facility onshore, and tow the installation to site floating on its own buoyancy. Bottom founded structure are lowered to the seabed by de-ballasting (see for instance Condeep or Cranefree), whilst floating structures are held in position with substantial mooring systems. The size of offshore lifts can be reduced by making the construction modular, with each module being constructed onshore and then lifted using a crane vessel into place onto the platform. A number of very large crane vessels were built in the 1970s which allow very large single modules weighing up to 14,000 tonnes to be fabricated and then lifted into place.

Specialist floating hotel vessels known as flotels are used to accommodate workers during the construction and hook-up phases. This is a high cost activity due to the limited space and access to materials. Oil platforms are key fixed installations from which drilling and production activity is carried out. Drilling rigs are either floating vessels for deeper water or jack-up designs which are a barge with liftable legs. Both of these types of vessel are constructed in marine yards but are often involved during the construction phase to pre-drill some production wells.

Other key factors in offshore construction are the weather window which defines periods of relatively light weather during which continuous construction or other offshore activity can take place. Safety is another key construction parameter, the main hazard obviously being a fall into the sea from which speedy recovery in cold waters is essential. The main types of vessels used for pipe laying are the "Derrick Barge (DB)", the "Pipelay Barge (LB)" and the "Derrick/Lay Barge (DLB)" combination. Diving Bells in offshore construction are mainly used in water depths greater than 120 ft, less than that, the divers use a metal basket driven from an "A" frame from the deck. The basket is lowered to the water level, then the divers enter the water from it to a maximum of 120 ft. Bells can go to 1500 ft, but are normally used at 400 to 800 ft. Offshore construction includes foundations engineering, structural design, construction, and/or repair of offshore structures, both commercial and military, including:

- Subsea oil and gas developments
- Offshore platforms - Fixed Platforms , Semi-submersibles, Spars, Tension leg platforms (TLPs), Floating Production Storage and Offloading (FPSOs), etc.
- Floating Oil and Gas Platforms - Semi-submersibles, Spars, TLPs, FPSOs, etc.

AMPELMANN SYSTEM

The Ampelmann system is a Dutch invention that makes it possible to transport personnel from a moving vessel to a platform offshore. The idea of the Ampelmann system was originated by Jan van der Tempel while visiting a conference in Berlin in 2002. After attending a presentation there on access to offshore wind turbines he noticed that current methods were ineffective and that there was a need for a better solution. Jan van der Tempel PhD invented the Ampelmann system while working on his thesis in offshore wind energy. In 2008, the system performed its first commercial task successfully.Workings

The Ampelmann system is a self-stabilizing platform that provides access from a moving vessel to a stable platform at sea. Six hydraulic legs work against any movement of the vessel caused by waves. The six legs are able to compensate any up-down, forward-backward, left-right and rotation movements. The length of the connecting gangway can be adjusted between 8.2 meters and 10 meters. It can be installed on location within 8 hours and operated by a single person. Each running system is operated and maintained by two employees of Ampelmann. These operators are specially trained mechanics who are in charge of a constant working system to offer a service with the highest safety standards.

CLIENTS

The customers that use this system are mainly operating in the offshore oil & gas industry and the offshore wind industry. They use the system to enable their employees to safely transfer to perform maintenance on offshore wind turbines or to work on an offshore oil rig. Both market segments are growing in a rapid pace. The offshore wind sector, for example, has grown 54% in 2009. This grow has mainly been caused by the shift to more sustainable forms of energy generation by governments all over the world. A good example of this is Germany that has plans to build more than forty wind parks in the near future. The customer relation of Ampelmann is only business-to-business (B2B). Platforms are usually owned by private companies, whereas wind parks are always run by private companies. There is no agent of independent status between Ampelmann and the user of the system.

FLOATING PRODUCTION STORAGE AND OFFLOADING

A floating production, storage and offloading (FPSO) unit is a floating vessel used by the offshore oil and gas industry for the processing of hydrocarbons and for storage of oil. An FPSO vessel is designed to receive hydrocarbons produced from nearby platforms or subsea template, process them, and store oil until it can be offloaded onto a tanker or, less frequently, transported through a pipeline.

FPSOs are preferred in frontier offshore regions as they are easy to install, and do not require a local pipeline infrastructure to export oil. FPSOs can be a conversion of an oil tanker or can be a vessel built specially for the application. A vessel used only to store oil (without processing it) is referred to as a floating storage and offloading vessel (FSO). There are also under construction (as at 2013) Floating Liquefied Natural Gas (FLNG) vessels, which will extract and liquify natural gas on board.

HISTORY

Oil has been produced from offshore locations since the late 1940s. Originally, all oil platforms sat on the seabed, but as exploration moved to deeper waters and more distant locations in the 1970s, floating production systems came to be used. The first oil FPSO was the Shell Castellon, built in Spain in 1977. Today, over 200 vessels are deployed worldwide as oil FPSOs.

In addition to the significant growth of this market sector, we are witnessing today the progressive extension of the significant knowledge base of building and operating these floating facilities to provide solutions for other segments of the oil and gas industry.

As an example, the Sanha LPG FPSO, which operates offshore Angola, is the first such vessel with complete onboard liquefied petroleum gas processing and export facilities. It can store up to 135,000 cubic meters of LPG while awaiting export tankers for offloading. Another very promising expansion is the progressive development of the floating LNG (FLNG) market. An LNG FPSO works under the same principles an oil FPSO works under, taking the well stream and separating out the natural gas (primarily methane and ethane) and producing LNG, which is stored and offloaded.

On July 29, 2009, Shell and Samsung announced an agreement to build up to 10 LNG FPSOs: Flex LNG has four contracts for smaller units at the same yard. As a very significant and momentous event, on May 20, 2011, Royal Dutch Shell announced the planned development of a Floating Liquefied

Natural Gas (FLNG) facility, which will be situated 200 km off the coast of Western Australia and is due for completion in around 2017. When it is finished, this will be the largest floating offshore facility. It will measure around 488m long and 74m wide, and when fully ballasted will weigh 600,000 tonnes. It will have a total storage capacity of 436,000 cubic metres of LNG, plus LPG condensate.

At the opposite (discharge and regasification) end of the LNG chain, the first ever conversion of an LNG carrier (Golar LNG owned Moss type LNG carrier) into an LNG floating storage and regasification unit was carried out in 2007 by Keppel shipyard in Singapore.

MECHANISMS

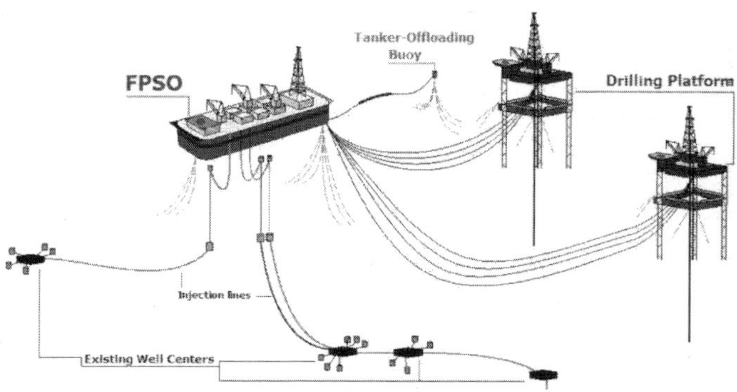

Fig. FPSO diagram

Oil produced from offshore production platforms can be transported to the mainland either by pipeline or by tanker. When a tanker is chosen to transport the oil, it is necessary to accumulate oil in some form of storage tank such that the oil tanker is not continuously occupied during oil production, and is only needed once sufficient oil has been produced to fill the tanker. At this point the transport tanker connects to the stern of the storage unit and offloads oil.

ADVANTAGES

Floating production, storage and offloading vessels are particularly effective in remote or deepwater locations where seabed pipelines are not cost effective. FPSOs eliminate the need to lay expensive long-distance pipelines from the processing facility to an onshore terminal. This can provide an economically attractive solution for smaller oil fields which can be exhausted in a few years and do not justify the expense of installing a pipeline. Furthermore, once the field is depleted, the FPSO can be moved to a new location.

SPECIFIC TYPES

A floating storage and offloading unit (FSO) is essentially a simplified FPSO without the capability for oil or gas processing. Most FSOs are converted single hull supertankers. An example is Knock Nevis, ex Seawise Giant, for many years the world's largest ship, which has been converted to an FSO for use offshore. The vessel was sold to Indian ship breakers, and renamed Mont for her final journey in December 2009. After clearing Indian customs, she was sailed to, and intentionally beached at Alang, Gujarat, India for demolition.

At the other end of the LNG logistics chain, where the natural gas is brought back to ambient temperature and pressure, specially modified ships may also be used as floating storage and regasification units (FSRUs). A LNG floating storage and regasification unit receives liquefied natural gas (LNG) from offloading LNG carriers, and the onboard regasification system provides natural gas exported to shore through risers and pipelines. Mooring systems for FSO, FPSO & FSU units are available in market which allow the vessel to be moored on an ice sheet.

VESSELS

Records

The FPSO operating in the deepest waters is the FPSO BW Pioneer, built and operated by BW Offshore on behalf of Petrobras Americas INC. The FPSO is moored at a depth of 2,600 m in Block 249 Walker Ridge in the US Gulf of Mexico and is rated for 100,000 bbl/d (16,000 m3/d). The EPCI contract was awarded in October 2007, and production started in early 2012. The FPSO conversion was carried out at Keppel Shipyard Tuas in Singapore, while the topsides were fabricated in modules at various international vendor locations. The FPSO has a disconnectable turret (APL). The vessel can disconnect in advance of hurricanes and reconnect with minimal down time.. A contract for an FPSO to operate in even deeper waters for Shell's Stones field in the US Gulf of Mexico was awarded to SBM Offshore in July 2013.

One of the world's largest FPSO is the Kizomba A, with a storage capacity of 2.2 million barrels (350,000 m3). Built at a cost of over US$800 million by Hyundai Heavy Industries in Ulsan, Korea, it is operated by Esso Exploration Angola (ExxonMobil). Located in 1200 meters (3,940 ft) of water at Deepwater block 200 statute miles (320 km) offshore in the Atlantic Ocean from Angola, Central Africa, it weighs 81,000 tonnes and is 285 meters long, 63 meters wide, and 32 meters high (935 ft by 207 ft (63 m) by 105 ft).

The first FSO in the Gulf of Mexico, The FSO Ta'Kuntah, has been in operation since August 1998. The FSO, owned and operated by MODEC, is under a service agreement with PEMEX Exploration and Production. The FSO Ta'Kuntah was installed as part of the Cantarell Field Development. The field is located in the Bay of Campeche, offshore Mexico's Yucatán peninsula. The FSO Ta'Kuntah is a converted ULCC tanker with a SOFEC external turret mooring system, two flexible risers connected in a lazy-S configuration between the turret and a pipeline end manifold (PLEM) on the seabed, and a unique offloading system. The FSO is designed to handle 800,000 bbl/d (130,000 m3/d) with no allowance for downtime. One of the world's smallest FPSO is the Crystal Ocean, operating in 137 m of water in the Bass Strait between Australia and Tasmania on the Basker Manta Field. It is leased by Roc Oil (Sydney-based international petroleum exploration and production company) from Rubicon Offshore and is operated on their behalf by AGR Asia Pacific; it is currently producing 5,000 bbl/d (790 m3/d).

The FPSO in the shallowest water depth of just 13 m is the Armada Perkasa in the Okoro field in Nigeria, West Africa, for Afren Energy. This spread moored (fixed orientation) vessel uses 100 mm, 150 mm and 200 mm bore DeepFlex non-steel flexible risers in a double lazy wave formation (with weights and distributed buoyancy) to accommodate the large motion offsets in an environment of extreme waves and currents. The Skarv FPSO, developed and engineered by Aker Solutions for BP Norge, will be the most advanced and largest FPSO deployed in the Norwegian Sea, offshore Mid Norway. Skarv is a gas condensate and oil field development. The development will tie in five sub-sea templates, and the FPSO has capacity to include several smaller wells nearby in the future. The process plant on the vessel can handle about 19,000,000 cubic metres per day (670,000,000 cu ft/d) of gas and 13,500 cubic metres per day (480,000 cu ft/d) of oil. An 80 km gas export pipe will tie in to Åsgard transport system. Aker Solutions (formerly Aker Kvaerner) developed the front-end design for the new floating production facility as well as the overall system design for the field and preparation for procurement and project management of the total field development.

The hull is an Aker Solutions proprietary "Tentechtm975" design. BP also selected Aker Solutions to perform the detail engineering, procurement and construction management assistance (EPcma) for the Skarv field development. The EPcma contract covers detail engineering and procurement work for the FPSO topsides as well as construction management assistance to BP including hull and topside facilities. The production start for the field is scheduled for August 2011. BP awarded the contract for fabrication of the Skarv FPSO hull to Samsung Heavy Industries in South Korea and the Turret contract to SBM. The FPSO has a length of 292m, breadth of 50.6m and is 29m deep and accommodate 100 people in single cabins. The hull will be delivered in January 2010.

INFRARED OPEN-PATH DETECTOR

Infrared open-path gas detectors send out a beam of infrared light, detecting gas anywhere along the path of the beam. This linear 'sensor' is typically a few metres up to a few hundred metres in length. Open-path detectors can be contrasted with Infrared point sensors. They are widely used in the petroleum and petrochemical industries, mostly to achieve very rapid gas leak detection for flammable gases at concentrations comparable to the lower flammable limit (typically a few percent by volume). They are also used, but so far to a lesser extent, in other industries where flammable concentrations can occur, such as in coal mining and water treatment. In principle the technique can also be used to detect toxic gases, for instance hydrogen sulfide, at the necessary parts-per-million concentrations, but the technical difficulties involved have so far prevented widespread adoption for toxic gases.

Usually, there are separate transmitter and receiver units at either end of a straight beam path. Alternatively, the source and receiver are combined, and the beam bounced off a retroreflector at the far end of the measurement path. For portable use, detectors have also been made which use the natural albedo of surrounding objects in place of the retroreflector. The presence of a chosen gas (or class of gases) is detected from its absorption of a suitable infrared wavelength in the beam. Rain, fog etc. in the measurement path can also reduce the strength of the received signal, so it is usual to make a simultaneous measurement at one or more reference wavelengths. The quantity of gas intercepted by the beam is then inferred from the ratio of the signal losses at the measurement and reference wavelengths. The calculation is typically carried out by a microprocessor which also carries out various checks to validate the measurement and prevent false alarms.

The measured quantity is the sum of all the gas along the path of the beam, sometimes termed the path-integral concentration of the gas. Thus the measurement has a natural bias (desirable in many applications) towards the total size of an unintentional gas release, rather than the concentration of the gas that has reached any particular point. Whereas the natural units of measurement for an Infrared point sensor are parts-per-million (ppm) or the percentage of the lower flammable limit (%LFL), the natural units of measurement for an open path detector are ppm.metres (ppmm) or LFL.metres (LFLm). For instance, the fire and gas safety system on an offshore platform in the North Sea typically has detectors set to a full-scale reading of 5LFLm, with low and high alarms triggered at 1LFLm and 3LFLm respectively.

ADVANTAGES AND DISADVANTAGES VERSUS FIXED-POINT DETECTORS

An open path detector usually costs more than a single point detector, so there is little incentive for applications that play to a point detector's strengths: where the point detector can be placed at the known location of the highest gas concentration, and a relatively slow response is acceptable. The open path detector excels in outdoor situations where, even if the likely source of the gas release is known, the evolution of the developing cloud or plume is unpredictable. Gas will almost certainly enter an extended linear beam before finding its way to any single chosen point. Also, point detectors in exposed outdoor locations require weather shields to be fitted, increasing the response time significantly. Open path detectors can also show a cost advantage in any application where a row of point detectors would be required to achieve the same coverage, for instance monitoring along a pipeline, or around the perimeter of a plant. Not only will one detector replace several, but the costs of installation, maintenance, cabling etc. are likely to be lower.

An important consideration for both types is a realistic assessment of their availability in service, distinguishing carefully between revealed and unrevealed downtime. The latter, where a detector appears to be working but is insensitive to gas, is especially serious. For instance, point detectors of the catalytic type are prone to poisoning by silicones and H2S, or by clogging of the gauze or sinter by water or ice. Point detectors of the infrared type are immune from the former but not the latter mechanism. Open path detectors suffer downtime from anything that blocks the path of the beam, such as people, vehicles or thick fog. However, they approach the ideal of eliminating unrevealed downtime because the loss of signal strength is easily made to generate a 'beam block' signal, distinct from zero gas or a hardware fault. Infrared detectors of both types, unlike passive sensors, invariably incorporate a microprocessor capable of self-checking the circuitry, so unrevealed downtime due to hardware failures is largely eliminated. The possibility of (revealed) downtime of an open path detector due to fog can be minimised by limiting the beam path to moderate lengths. In fog-prone areas, such as the North Sea, paths are often limited to 15 to 20 metres for this reason, even though the detectors are capable of much greater distances in clear air.

COMPONENT PARTS

In principle any source of infrared radiation could be used, together with an optical system of lenses or mirrors to form the transmitted beam. In practice the following sources have been used, always with some form of modulation to aid the signal processing at the receiver:

- An incandescent light bulb, modulated by pulsing the current powering the filament or by a mechanical chopper. For systems used outdoors, it is difficult for an incandescent source to compete with the intensity of sunlight when the sun shines directly into the receiver. Also, it is difficult to achieve modulation frequencies distinguishable from those that can be produced naturally, for instance by heat shimmer or by sunlight reflecting off waves at sea.
- A gas-discharge lamp is capable of exceeding the spectral power of direct sunlight in the infrared, especially when pulsed. Modern open path systems typically use a xenon flashtube powered by a capacitor discharge. Such pulsed sources are inherently modulated.

- A semiconductor laser provides a relatively weak source, but one that can be modulated at high frequency in wavelength as well as amplitude. This property permits various signal processing schemes based on Fourier analysis, of use when the absorption of the gas is weak but narrow in spectral linewidth.

The precise wavelength passbands used must be isolated from the broad infrared spectrum. In principle any conventional spectrometer technique is possible, but the NDIR technique with multilayer dielectric filters and beamsplitters is most often used. These wavelength-defining components are usually located in the receiver, although one design has shared the task with the transmitter.

At the receiver, the infrared signal strengths are measured by some form of infrared detector. Generally photodiode detectors are preferred, and are essential for the higher modulation frequencies, whereas slower photoconductive detectors may be required for longer wavelength regions. The signals are fed to low-noise amplifiers, then invariably subject to some form of digital signal processing. The absorption coefficient of the gas will vary across the passband, so the simple Beer-Lambert law cannot be applied directly. For this reason the processing usually employs a calibration table, applicable for a particular gas, type of gas, or gas mixture, and sometimes configurable by the user.

OPERATING WAVELENGTHS

The choice of infrared wavelengths used for the measurement largely defines the detector's suitability for a particular applications. Not only must the target gas (or gases) have a suitable absorption spectrum, the wavelengths must lie within a spectral window so the air in the beam path is itself transparent. These wavelength regions have been used:

- 3.4 pm region. All hydrocarbons and their derivatives absorb strongly, due to the C-H stretch mode of molecular vibration. It is commonly used in infrared point detectors where path lengths are necessarily short, and for open-path detectors requiring parts-per-million sensitivity. A disadvantage for many applications is that methane absorbs relatively weakly compared to heavier hydrocarbons, leading to large inconsistencies of calibration. For open-path detection of flammable concentrations the absorption for non-methane hydrocarbons is so strong that the measurement saturates, a significant gas cloud appearing 'black'. This wavelength region is beyond the transmission range of borosilicate glass, so windows and lenses must be made of more expensive materials and tend to be small in aperture.
- 2.3 pm region. All hydrocarbons and their derivatives have absorption coefficients appropriate for open path detection at

flammable concentrations. A useful advantage in practical applications is that the detector's response to many different gases and vapours is relatively uniform when expressed in terms of the lower flammable limit. Borosilicate glass retains useful transmission in this wavelength region, allowing large aperture optics to be produced at moderate cost.

- 1.6 pm region. A wide range of gases absorb in the near-infrared. Typically the absorption coefficients are relatively weak, but light molecules show narrow, individually resolved spectral lines rather than broad bands. This results in relatively large values of the gradient and curvature of the absorption with respect to wavelength, enabling semiconductor laser-based systems to distinguish gas molecules very specifically; for instance hydrogen sulfide, or methane to the exclusion of heavier hydrocarbons.

HISTORY

The first open-path detector offered for routine industrial use, as distinct from research instruments built in small numbers, was the Wright and Wright 'Pathwatch' in the US, 1983. Acquired by Det-tronics in 1992, the detector operated in the 3.4 pm region with a powerful incandescent source and a mechanical chopper. It did not achieved large volume sales, mainly because of cost and doubts about long-term reliability with moving parts. Beginning in 1985, Shell Research in UK was funded by Shell Natural Gas to develop an open-path detector with no moving parts.

The advantages of the 2.3 pm wavelength were identified, and a research prototype was demonstrated. This design had a combined transmitter-receiver with a corner-cube retroreflector at 50 m. It used a pulsed incandescent lamp, PbS photoconductive detectors in the gas and reference channels, and an Intel 8031 microprocessor for signal processing. In 1987 Shell licenced this technology to Sieger-Zellweger (later Honeywell) who designed and marketed their industrial version as the 'Searchline', using a retro-reflective panel made up of multiple corner-cubes.

This was the first open-path detector to be certified for use in hazardous areas and to have no moving parts. Later work by Shell Research used two alternately pulsed incandescent sources in the transmitter and a single PbS detectors in the receiver, avoiding zero drifts caused by the variable responsivity of PbS detectors. This technology was offered to Sieger-Zellweger, and later licensed to PLMS. a company part-owned by Shell Ventures UK. The PLMS GD4001/2 in 1991 were the first detectors to achieve a truly stable zero without moving parts or software compensation of slow drifts. They were also the first infrared gas detectors of any kind to be certified intrinsically safe. The Israeli company Spectronix (also Spectrex) made an important advance in 1996 with their SafEye, the first to use a flash tube source, followed by Sieger-Zellweger with their Searchline Excel in 1998. In 2001 the PLMS Pulsar, soon afterwards acquired by Dräger as their Polytron Pulsar, was the first detector to incorporate sensing to monitor the mutual alignment of the transmitter and receiver during both installation and routine operation.

OIL PLATFORM

An oil platform, (offshore platform or colloquially oil rig) is a large structure with facilities to drill wells, to extract and process oil and natural gas, and to temporarily store product until it can be brought to shore for refining and marketing. In many cases, the platform contains facilities to house the workforce as well.

Depending on the circumstances, the platform may be fixed to the ocean floor, may consist of an artificial island, or may float. Remote subsea wells may also be connected to a platform by flow lines and by umbilical connections. These subsea solutions may consist of one or more subsea wells, or of one or more manifold centres for multiple wells.

HISTORY

Around 1891, the first submerged oil wells were drilled from platforms built on piles in the fresh waters of the Grand Lake St. Marys (a.k.a. Mercer County Reservoir) in Ohio. The wide but shallow reservoir was built from 1837 to 1845 to provide water to the Miami and Erie Canal. Around 1896, the first submerged oil wells in salt water were drilled in the portion of the Summerland field extending under the Santa Barbara Channel in California. The wells were drilled from piers extending from land out into the channel.

Other notable early submerged drilling activities occurred on the Canadian side of Lake Erie in the 1900s and Caddo Lake in Louisiana in the 1910s. Shortly thereafter, wells were drilled in tidal zones along the Gulf Coast of Texas and Louisiana. The Goose Creek field near Baytown, Texas is one such example. In the 1920s, drilling was done from concrete platforms in Lake Maracaibo, Venezuela.

The oldest subsea well recorded in Infield's offshore database is the Bibi Eibat well which came on stream in 1923 in Azerbaijan. Landfill was used to raise shallow portions of the Caspian Sea. In the early 1930s, the Texas Company developed the first mobile steel barges for drilling in the brackish coastal areas of the gulf.

In 1937, Pure Oil Company (now part of Chevron Corporation) and its partner Superior Oil Company (now part of ExxonMobil Corporation) used a fixed platform to develop a field in 14 feet (4.3 m) of water, one mile (1.6 km) offshore of Calcasieu Parish, Louisiana. In 1946, Magnolia Petroleum Company (now part of ExxonMobil) erected a drilling platform in 18 ft (5.5 m) of water, 18 miles off the coast of St. Mary Parish, Louisiana. In early 1947, Superior Oil erected a drilling/production platform in 20 ft (6.1 m) of water some 18 miles off Vermilion Parish, Louisiana. But it was Kerr-McGee Oil Industries (now Anadarko Petroleum Corporation), as operator for partners Phillips Petroleum (ConocoPhillips) and Stanolind Oil & Gas (BP), that completed its historic Ship Shoal Block 32 well in October 1947, months before Superior actually drilled a discovery from their Vermilion platform farther offshore. In any case, that made Kerr-McGee's well the first oil discovery drilled out of sight of land.

The Thames Sea Forts of World War II are considered the direct predecessors of modern offshore platforms. Having been pre-constructed in a very short time, they were then floated to their location and placed on the shallow bottom of the Thames estuary.

TYPES

Larger lake- and sea-based offshore platforms and drilling rigs are some of the largest moveable man-made structures in the world. There are several types of oil platforms and rigs:

1, 2) conventional fixed platforms; 3) compliant tower; 4, 5) vertically moored tension leg and mini-tension leg platform; 6) Spar; 7,8) Semi-submersibles; 9) Floating production, storage, and offloading facility; 10) sub-sea completion and tie-back to host facility.

Fixed platforms

These platforms are built on concrete or steel legs, or both, anchored directly onto the seabed, supporting a deck with space for drilling rigs, production facilities and crew quarters. Such platforms are, by virtue of their immobility, designed for very long term use (for instance the Hibernia platform). Various types of structure are used, steel jacket, concrete caisson, floating steel and even floating concrete. Steel jackets are vertical sections made of tubular steel members, and are usually piled into the seabed. To see more details regarding Design, construction and installation of such platforms refer to: and .

Concrete caisson structures, pioneered by the Condeep concept, often have in-built oil storage in tanks below the sea surface and these tanks were often used as a flotation capability, allowing them to be built close to shore (Norwegian fjords and Scottish firths are popular because they are sheltered and deep enough) and then floated to their final position where they are sunk to the seabed. Fixed platforms are economically feasible for installation in water depths up to about 520 m (1,710 ft).

Compliant towers

These platforms consist of slender flexible towers and a pile foundation supporting a conventional deck for drilling and production operations. Compliant towers are designed to sustain significant lateral deflections and forces, and are typically used in water depths ranging from 370 to 910 metres (1,210 to 2,990 ft).

Semi-submersible platform

These platforms have hulls (columns and pontoons) of sufficient buoyancy to cause the structure to float, but of weight sufficient to keep the structure upright. Semi-submersible platforms can be moved from place to place; can be ballasted up or down by altering the amount of flooding in buoyancy tanks; they are generally anchored by combinations of chain, wire rope or polyester rope, or both, during drilling or production operations, or both, though they can also be kept in place by the use of dynamic positioning. Semi-submersibles can be used in water depths from 60 to 3,000 metres (200 to 10,000 ft).

Jack-up drilling rigs

Jack-up Mobile Drilling Units (or jack-ups), as the name suggests, are rigs that can be jacked up above the sea using legs that can be lowered, much like jacks. These MODUs (Mobile Offshore Drilling Units) are typically used in water depths up to 120 metres (390 ft), although some designs can go to 170 m (560 ft) depth. They are designed to move from place to place, and then anchor themselves by deploying the legs to the ocean bottom using a rack and pinion gear system on each leg.

Drillships

A drillship is a maritime vessel that has been fitted with drilling apparatus. It is most often used for exploratory drilling of new oil or gas wells in deep water but can also be used for scientific drilling. Early versions were built on a modified tanker hull, but purpose-built designs are used today. Most drillships are outfitted with a dynamic positioning system to maintain position over the well. They can drill in water depths up to 3,700 m (12,100 ft).

Floating production systems

The main types of floating production systems are FPSO (floating production, storage, and offloading system). FPSOs consist of large monohull structures, generally (but not always) shipshaped, equipped with processing facilities. These platforms are moored to a location for extended periods, and do not actually drill for oil or gas. Some variants of these applications, called FSO (floating storage and offloading system) or FSU (floating storage unit), are used exclusively for storage purposes, and host very little process equipment. This is one of the best sources for having floating production. The world's first Floating Liquefied Natural Gas (FLNG) facility is currently under development. See the section on Particularly large examples below.

Tension-leg platform

TLPs are floating platforms tethered to the seabed in a manner that eliminates most vertical movement of the structure. TLPs are used in water depths up to about 2,000 meters (6,000 ft). The "conventional" TLP is a 4-column design which looks similar to a semisubmersible. Proprietary versions include the Seastar and MOSES mini TLPs; they are relatively low cost, used in water depths between 180 and 1,300 metres (590 and 4,300 ft). Mini TLPs can also be used as utility, satellite or early production platforms for larger deepwater discoveries.

Gravity-based structure

A GBS can either be steel or concrete and is usually anchored directly onto the seabed. Steel GBS are predominantly used when there is no or limited availability of crane barges to install a conventional fixed offshore platform, for example in the Caspian Sea. There are several steel GBS in the world today (e.g. offshore Turkmenistan Waters (Caspian Sea) and offshore New Zealand). Steel GBS do not usually provide hydrocarbon storage capability.

It is mainly installed by pulling it off the yard, by either wet-tow or/and dry-tow, and self-installing by controlled ballasting of the compartments with sea water. To position the GBS during installation, the GBS may be connected to either a transportation barge or any other barge (provided it is large enough to support the GBS) using strand jacks. The jacks shall be released gradually whilst the GBS is ballasted to ensure that the GBS does not sway too much from target location.

Spar platforms

Spars are moored to the seabed like TLPs, but whereas a TLP has vertical tension tethers, a spar has more conventional mooring lines. Spars have to-date been designed in three configurations: the "conventional" one-piece cylindrical hull, the "truss spar" where the midsection is composed of truss elements connecting the upper buoyant hull (called a hard tank) with the bottom soft tank containing permanent ballast, and the "cell spar" which is built from multiple vertical cylinders.

The spar has more inherent stability than a TLP since it has a large counterweight at the bottom and does not depend on the mooring to hold it upright. It also has the ability, by adjusting the mooring line tensions (using chain-jacks attached to the mooring lines), to move horizontally and to position itself over wells at some distance from the main platform location. The first production spar was Kerr-McGee's Neptune, anchored in 590 m (1,940 ft) in the Gulf of Mexico; however, spars (such as Brent Spar) were previously used as FSOs.

Eni's Devil's Tower located in 1,710 m (5,610 ft) of water, in the Gulf of Mexico, was the world's deepest spar until 2010. The world's deepest platform is currently the Perdido spar in the Gulf of Mexico, floating in 2,438 meters of water. It is operated by Royal Dutch Shell and was built at a cost of $3 billion.

The first truss spars were Kerr-McGee's Boomvang and Nansen. The first (and only) cell spar is Kerr-McGee's Red Hawk.

Condeep platforms

Condeep platform got his breakthrough summer 1973. When a contract was signed between the Norwegian Contractors (NC) and Mobil on the construction of Beryl A and the Shell Brent B. It had not been implemented so bold building projects in Norway since Rjukan development. The Norwegian Condeep platforms attracted international attention, and construction was awarded the prize for technological innovation in the world's largest oil conference in Houston in 1975.

Internationally Condeep epitome of oil operations in the North Sea. Condeep platforms have been the most significant independent contributions to the offshore industry. The latest and greatest Condeep platform, Troll A, which was submitted to Shell in 1995. The first concrete platforms in the UK sector was Condeep platforms built by the NC. Several of the next British concrete platform had multiply design resembling Condeep. More about other concrete facilities found in chapter 3.3. In 1978, there were demands from NPD that platforms should also be designed for removal, so this is included for the nine Condeep platforms from the Statfjord B. However, the full extent of the removal process is not generally considered in the original design.

Normally unmanned installations (NUI)

These installations (sometimes called toadstools) are small platforms, consisting of little more than a well bay, helipad and emergency shelter. They are designed to be operated remotely under normal conditions, only to be visited occasionally for routine maintenance or well work.

Conductor support systems

These installations, also known as satellite platforms, are small unmanned platforms consisting of little more than a well bay and a small process plant. They are designed to operate in conjunction with a static production platform which is connected to the platform by flow lines or by umbilical cable, or both.

Particularly large examples

The Petronius Platform is a compliant tower in the Gulf of Mexico modeled after the Hess Baldpate platform, which stands 2,000 feet (610 m) above the ocean floor. It is one of the world's tallest structures. The Hibernia platform in Canada is the world's largest (in terms of weight) offshore platform, located on the Jeanne D'Arc Basin, in the Atlantic Ocean off the coast of Newfoundland. This gravity base structure (GBS), which sits on the ocean floor, is 111 metres (364 ft) high and has storage capacity for 1.3 million barrels (210,000 m3) of crude oil in its 85-metre (279 ft) high caisson. The platform acts as a small concrete island with serrated outer edges designed to withstand the impact of an iceberg. The GBS contains production storage tanks and the remainder of the void space is filled with ballast with the entire structure weighing in at 1.2 million tons.

Royal Dutch Shell is currently developing the first Floating Liquefied Natural Gas (FLNG) facility, which will be situated approximately 200km off the coast of Western Australia and is due for completion around 2017. When finished, it will be the largest floating offshore facility. It is expected to be approximately 488m long and 74m wide with displacement of around 600,000t when fully ballasted.

MAINTENANCE AND SUPPLY

A typical oil production platform is self-sufficient in energy and water needs, housing electrical generation, water desalinators and all of the equipment necessary to process oil and gas such that it can be either delivered directly onshore by pipeline or to a floating platform or tanker loading facility, or both. Elements in the oil/gas production process include wellhead, production manifold, production separator, glycol process to dry gas, gas compressors, water injection pumps, oil/gas export metering and main oil line pumps.

Larger platforms assisted by smaller ESVs (emergency support vessels) like the British Iolair that are summoned when something has gone wrong, e.g. when a search and rescue operation is required. During normal operations, PSVs (platform supply vessels) keep the platforms provisioned and supplied, and AHTS vessels can also supply them, as well as tow them to location and serve as standby rescue and firefighting vessels.

CREW

Essential personnel

Not all of the following personnel are present on every platform. On smaller platforms, one worker can perform a number of different jobs. The following also are not names officially recognized in the industry:

- OIM (offshore installation manager) who is the ultimate authority during his/her shift and makes the essential decisions regarding the operation of the platform;

- operations team leader (OTL);
- offshore operations engineer (OOE) who is the senior technical authority on the platform;
- PSTL or operations coordinator for managing crew changes;
- dynamic positioning operator, navigation, ship or vessel maneuvering (MODU), station keeping, fire and gas systems operations in the event of incident;
- automation systems specialist, to configure, maintain and troubleshoot the process control systems (DCS), process safety systems, emergency support systems and vessel management systems;
- second mate to meet manning requirements of flag state, operates fast rescue craft, cargo operations, fire team leader;
- third mate to meet manning requirements of flag state, operate fast rescue craft, cargo operations, fire team leader;
- ballast control operator to operate fire and gas systems;
- crane operators to operate the cranes for lifting cargo around the platform and between boats;
- scaffolders to rig up scaffolding for when it is required for workers to work at height;
- coxswains to maintain the lifeboats and manning them if necessary;
- control room operators, especially FPSO or production platforms;
- catering crew, including people tasked with performing essential functions such as cooking, laundry and cleaning the accommodation;
- production techs to run the production plant;
- helicopter pilot(s) living on some platforms that have a helicopter based offshore and transporting workers to other platforms or to shore on crew changes;
- maintenance technicians (instrument, electrical or mechanical).

Incidental personnel

Drill crew will be on board if the installation is performing drilling operations. A drill crew will normally comprise:

- Toolpusher
- Driller
- Roughnecks
- Roustabouts
- Company man
- Mud engineer
- Derrickhand
- Geologist
- Welders and Welder Helpers

Well services crew will be on board for well work. The crew will normally comprise:

- Well services supervisor
- Wireline or coiled tubing operators
- Pump operator
- Pump hanger and Ranger

DRAWBACKS

Risks

The nature of their operation - extraction of volatile substances sometimes under extreme pressure in a hostile environment - means risk; accidents and tragedies occur regularly. The U.S. Minerals Management Service reported 69 offshore deaths, 1,349 injuries, and 858 fires and explosions on offshore rigs in the Gulf of Mexico from 2001 to 2010. In July 1988, 167 people died when Occidental Petroleum's Piper Alpha offshore production platform, on the Piper field in the UK sector of the North Sea, exploded after a gas leak. The resulting investigation conducted by Lord Cullen and publicized in the first Cullen Report was highly critical of a number of areas, including, but not limited to, management within the company, the design of the structure, and the Permit to Work System. The report was commissioned in 1988, and was delivered November 1990. The accident greatly accelerated the practice of providing living accommodations on separate platforms, away from those used for extraction.

However, this was in itself a hazardous environment. In March 1980, the 'flotel' (floating hotel) platform Alexander L. Kielland capsized in a storm in the North Sea with the loss of 123 lives. In 2001, Petrobras 36 in Brazil exploded and sank five days later, killing 11 people.

Given the number of grievances and conspiracy theories that involve the oil business, and the importance of gas/oil platforms to the economy, platforms in the United States are believed to be potential terrorist targets. Agencies and military units responsible for maritime counter-terrorism in the US (Coast Guard, Navy SEALs, Marine Recon) often train for platform raids. On April 21, 2010, the Deepwater Horizon platform, 52 miles off-shore of Venice, Louisiana, (property of Transocean and leased to BP) exploded, killing 11 people, and sank two days later. The resulting undersea gusher, conservatively estimated to exceed 20 million US gallons (76,000 m3) as of early June, 2010, became the worst oil spill in US history, eclipsing the Exxon Valdez oil spill.

ECOLOGICAL EFFECTS

In British waters, the cost of removing all platform rig structures entirely was estimated in 1995 at £1.5 billion, and the cost of removing all structures including pipelines-called a "clean sea" approach-at £3 billion.

In the United States, Marine Biologist Milton Love has proposed that oil platforms off the California coast be retained as artificial reefs, instead of being dismantled (at great cost), because he has found them to be havens for many of the species of fish which are otherwise declining in the region, in the course of 11 years of research. Love is funded mainly by government agencies, but also in small part by the California Artificial Reef Enhancement Program. NOAA has said it is considering this course of action, but wants money to study the effects of the rigs in detail. Divers have been used to assess the fish populations surrounding the platforms. In the Gulf of Mexico, more than 140 platforms have been similarly converted.

DEEPEST OIL PLATFORMS

The world's deepest oil platform is the floating Perdido which is a spar platform in the Gulf of Mexico in a water depth of 2,438 metres (7,999 ft).

Non-floating compliant towers and fixed platforms, by water depth:

- Petronius Platform, 535 m (1,755 ft)
- Baldpate Platform, 502 m (1,647 ft)
- Bullwinkle Platform, 413 m (1,355 ft)
- Pompano Platform, 393 m (1,289 ft)
- Benguela-Belize Lobito-Tomboco Platform, 390 m (1,280 ft)
- Tombua Landana Platform, 366 m (1,201 ft)
- Harmony Platform, 366 m (1,201 ft)
- Troll A Platform, 303 m (994 ft)
- Gulfaks C Platform, 217 m (712 ft)

SUCTION CAISSON

Suction caissons (also referred to as suction anchors, suction piles or suction buckets) are a new form of offshore foundation that have a number of advantages over conventional offshore foundations, mainly being quicker to install than piles and being easier to remove during decommissioning. Suction caissons are now used extensively worldwide for anchoring large offshore installations to the seafloor at great depths.

Oil and gas recovery at great depth could have been a very difficult task without the suction anchor technology, which was developed and used for the first time in the North Sea 30 years ago. The use of suction caissons/anchors has now become common practice worldwide. Statistics from 2002 revealed that 485 suction caissons had been installed in more than 50 different localities around the world, in depths to about 2000 m.

Suction caissons have been installed in most of the deep water oil producing areas around the world: The North Sea, Gulf of Mexico, offshore West Africa, offshore Brazil, West of Shetland, South China Sea, Adriatic Sea and Timor Sea. No reliable statistics has been produced after 2002, but the use of suction caissons is still rising.

DESCRIPTION

A suction caisson can effectively be described as an upturned bucket that is embedded in the marine sediment. This embedment is either achieved through pushing or by creating a negative pressure inside the caisson skirt: both of these techniques have the effect of securing the caisson into the sea bed. The foundation can also be rapidly removed by reversing the installation process, applying an overpressure inside the caisson skirt.

The concept of suction technology was developed for projects where gravity loading is not sufficient for pressing foundation skirts into the ground. The technology was also developed for anchors subject to large tension forces due to waves and stormy weather. The suction caisson technology functions very well in a seabed with soft clays or other low strength sediments. The suction caissons are in many cases easier to install than piles, which must be driven (hammered) into the ground.

Mooring lines are usually attached to the side of the suction caisson at the optimal load attachment point, which must be calculated for each caisson. Once installed, the caisson acts much like a short rigid pile and is capable of resisting both lateral and axial loads. Limit equilibrium methods or 3D finite element analyses are used to calculate the holding capacity.

HISTORY

Suction caissons were first used as anchors for floating structures in the offshore oil and gas industry, including offshore platforms such as the Draupner E oil rig. They have also been proposed for use in conjunction with offshore wind turbines.

There are great differences between the first small suction caissons that were installed for Shell at the Gorm field in the North Sea in 1981 and the large suction caissons that were installed for the Diana platform in the Gulf of Mexico in 1999. The twelve suction caissons on the Gorm field were intended to secure a simple loading buoy device at a depth of 40 metres, while the installation of suction anchors for the Diana platform was a world record in itself at that time, concerning water depth and size of anchors. The height of the Diana suction caissons is 30 metres, their diameter 6.5 metres, and they were installed at a depth of about 1500 m on soft clay deposits. Since then, suction caissons has been installed at even larger depths, but the Diana installation was a technology breakthrough for the 20th century.

An important development step for the suction caisson technology emerged from cooperation between the former operator in the North Sea, Saga Petroleum AS, and the Norwegian Geotechnical Institute (NGI). Saga Petroleum's oil-producing Snorre A platform was a tension-leg platform of a type that in other parts of the world would have been founded with up to 90 metres long piles. Unfortunately on the Snorre oil field, it was difficult to use long piles due to the presence of huge pebbles at 60 m depth under the seabed.

Saga Petroleum decided therefore to use suction caissons, which were analysed by NGI. These analyses were verified from extensive model tests. The calculations showed that the platform could be safely secured by suction caissons of only 12 m in length. Snorre A started to produce oil in 1992 and is now operated by the Norwegian oil company Statoil.

GRAVITY OIL PLATFORMS

Suction caissons have a lot of similarities with foundation design principles and solutions for the big gravity oil platforms that were installed in the North Sea when the offshore oil production started there in the beginning of 1970's. The first gravity oil platform on the Ekofisk oil field had a foundation area that was as big as a football field, and it was placed on a seabed with very dense sand. The platform was designed to tolerate waves up to 24 m in height.

As the installation of oil platforms continued in the North Sea, in areas with poor ground conditions such as soft clays, they were designed to survive even higher storm waves. These platforms were founded on a system of cylindrical skirts that were penetrated into the ground under combined gravity load and underpressure. The oil platform at the Gullfaks C field was equipped with 22 m long skirts. The Troll A platform is founded in 330 m depth with 30 m long skirts and is the world's biggest gravity platform.

RESEARCH AND DEVELOPMENT

The Norwegian Geotechnical Institute (NGI) has been heavily involved with the concept development, design and installation of suction anchors from the start. The project "Application of offshore bucket foundations and anchors in lieu of conventional designs" (1994-1998) was sponsored by 15 international petroleum and industry companies and was one of the most important studies. The project "Skirted foundations and anchors in clay" (1997-1999) was sponsored by 19 international companies organized through the Offshore Technology Research Center (OTRC) in the USA, and the project "Skirted offshore foundations and anchors in sand" (1997-2000) was sponsored by 8 international companies. The main conclusions from the projects were presented in the 1999 OTC paper no 10824.

An industry sponsored study on the design and analysis of deepwater anchors in soft clay was completed in 2003, where NGI participated together with OTRC and Centre for Offshore Foundation Systems (COFS) in Australia. The overall objective was to provide the API Geotechnical Workgroup (RG7) and the Deepstar Joint Industry Project VI with background, data and other information needed to develop a widely applicable recommended practice for the design and installation of deepwater anchors. The Norwegian classification society DNV (Det Norske Veritas), active worldwide in risk analysis and safety evaluation of special constructions, has produced a recommended practice report on the design procedures for suction anchors which is based on close cooperation with NGI. The main information from the project was presented in the 2006 OTC paper no 18038.

In 2002 NGI established the subsidiary NGI Inc in Houston. The subsidiary has since been awarded the detailed geotechnical design for more than 15 suction anchor projects in the Gulf of Mexico, and among these the challenging Mad Dog Spar project involving design of anchors located in old slide deposits below the Sigsbee Escarpment. For further information reference can be made to the 2006 OTC papers no 17949 and 17950.

5

Mooring (Watercraft)

A mooring refers to any permanent structure to which a vessel may be secured. Examples include quays, wharfs, jetties, piers, anchor buoys, and mooring buoys. A ship is secured to a mooring to forestall free movement of the ship on the water. An anchor mooring fixes a vessel's position relative to a point on the bottom of a waterway without connecting the vessel to shore. As a verb, mooring refers to the act of attaching a vessel to a mooring.

The term probably stems from the Dutch verb meren (to moor), which has been used in English since the end of the 15th century.

PERMANENT ANCHOR MOORING

These moorings are used instead of temporary anchors because they have considerably more holding power, cause less damage to the marine environment, and are convenient. They are also occasionally used to hold floating docks in place. There are several kinds of moorings:

SWING MOORINGS

Swing moorings also known as simple or single-point moorings, are the simplest and most common kind of mooring. A swing mooring consists of a single anchor at the bottom of a waterway with a rode (a rope, cable, or chain) running to a float on the surface. The float allows a vessel to find the rode and connect to the anchor. These anchors are known as swing moorings because a vessel attached to this kind of mooring swings in a circle when the direction of wind or tide changes.

FORE AND AFT MOORINGS

Fore and aft moorings, also known as double moorings or twin moorings, are just a pair of swing moorings with an additional rode between the two primary rodes to remove confusion as to which moorings are paired. A fore and aft mooring may have one or two floats. Fore and aft moorings fix a vessel's position more precisely than a swing mooring, and therefore allow a much greater density of vessels to be moored. They are used in particularly congested harbours, like those of Santa Catalina Island, California.

PILE MOORINGS

Pile moorings are poles driven into the bottom of the waterway with their tops above the water. Vessels then tie mooring lines to two or four piles to fix their position between those piles. Pile moorings are common in New Zealand but rare elsewhere.

While many mooring buoys are privately owned, some are available for public use. For example, on the Great Barrier Reef off the Australian coast, a vast number of public moorings are set out in popular areas where boats can moor. This is to avoid the massive damage that would be caused by many vessels anchoring.

There are four basic types of permanent anchors used in moorings:

- Dead weights are the simplest type of anchor. They are generally made as a large concrete block with a rode attached which resists movement with sheer weight; and, to a small degree, by settling into the substrate. The advantages are that they are simple and cheap. A dead weight mooring that drags in a storm still holds well in its new position. Such moorings are better suited to rocky bottoms where other mooring systems do not hold well. The disadvantages are that they are heavy, bulky, and awkward.
- Mushroom anchors are the most common anchors and work best for softer seabeds such as mud, sand, or silt. They are shaped like an upside-down mushroom which can be easily buried in mud or silt. The advantage is that it has up to ten times the holding-power-to-weight ratio compared to a dead weight mooring; disadvantages include high cost, limited success on rocky or pebbly substrates, and the long time it takes to reach full holding capacity.
- Screw-in moorings are a modern method. The anchor in a screw-in mooring is a shaft with wide blades spiraling around it so that it can be screwed into the substrate. The advantages include high holding-power-to-weight ratio and small size (and thus relative cheapness). The disadvantage is that a diver is usually needed to install, inspect, and maintain these moorings.
- Multiple anchor mooring systems use two or more (often three) light weight temporary-style anchors set in an equilateral arrangement and all chained to a common center from which a conventional rode extends to a mooring buoy. The advantages are minimized mass, ease of deployment, high holding-power-to-weight ratio, and availability of temporary-style anchors.

MOORING TO A SHORE FIXTURE

A vessel can be made fast to any variety of shore fixtures from trees and rocks to specially constructed areas such as piers and quays. The word pier is used in the following explanation in a generic sense.

Mooring is often accomplished using thick ropes called mooring lines or hawsers. The lines are fixed to deck fittings on the vessel at one end, and fittings on the shore, such as bollards, rings, or cleats, on the other end. Mooring requires cooperation between people on the pier and on a vessel. For larger vessels, heavy mooring lines are often passed to the people on the shore by use of smaller, weighted heaving lines. Once the mooring line is attached to the bollard, it is pulled tight. On large ships, this tightening can be accomplished with the help of heavy machinery called mooring winches or capstans.

For the heaviest cargo ships, more than a dozen mooring lines can be required. Small vessels generally take 4 to 6 mooring lines. Mooring lines are usually made out of synthetic materials such as nylon. Nylon is easy to work with and lasts for years, but has a property of very great elasticity. This elasticity has its advantages and disadvantages.

The main advantage is that during an event, such as a high wind or the close passing of another ship, excess stress can be spread among several lines On the other hand, if a highly stressed nylon line does break, or part, it causes a very dangerous phenomenon called "snapback" which can cause fatal injuries.

Snapback is analogous to stretching a rubber band to its breaking point between the hands, and then suffering a stinging blow from the retracting loose ends of the band - in the case of a heavy mooring line this blow carries much more force and can inflict severe injuries or sever limbs. Mooring lines made from materials such as Dyneema and Kevlar have much less elasticity and therefore much safer to use, but the lines do not float on the water, and tend to sink, are costly, so they are used less frequently. Manila rope is preferred.

Some ships use wire rope for one or more of their mooring lines. Wire rope is hard to handle and maintain. There is also a risk of using wire rope on a ship's stern in the vicinity of its propeller. Combination mooring lines made of both wire rope and synthetic line can also be used. This results in a hawser. This is more elastic and easier to handle than a wire rope, but not as elastic as a pure synthetic line. Special safety precautions must be followed when constructing a combination mooring line.

A typical mooring scheme

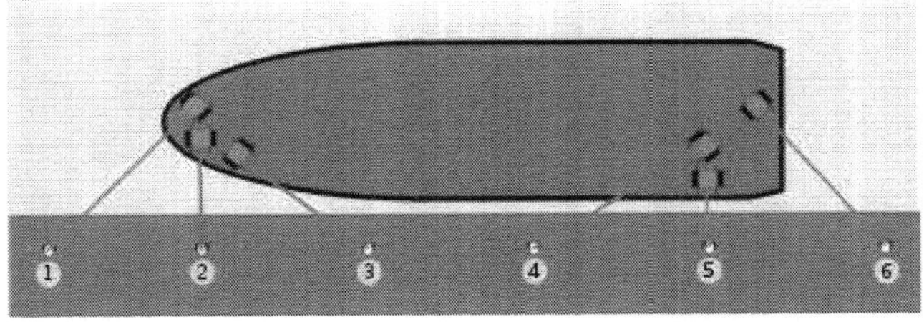

Number	Name	Purpose
1	Bow line	Prevent backwards movement
2	Forward Breast line	Keep close to pier
3	After Bow Spring line	Prevent from advancing
4	Forward Quarter Spring line	Prevent from moving back
5	Quarter Breast line	Keep close to pier
6	Stern line	Prevent forwards movement

The two-headed mooring bitt is a fitting often-used in mooring. The rope is hauled over the bitt, pulling the vessel toward the bitt. In the second step, the rope is tied to the bitt, as shown. This tie can be put and released very quickly. In quiet conditions, such as on a lake, one person can moor a 260-tonne ship in just a few minutes.

The basic rode system is a line, cable, or chain several times longer than the depth of the water running from the anchor to the mooring buoy, the longer the rode is the shallower the angle of force on the anchor (it has more scope). A shallower scope means more of the force is pulling horizontally so that ploughing into the substrate adds holding power but also increases the swinging circle of each mooring, so lowering the density of any given mooring field. By adding weight to the bottom of the rode, such as the use of a length of heavy chain, the angle of force can be dropped further. Unfortunately, this scrapes up the substrate in a circular area around the anchor. A buoy can be added along the lower portion of rode to hold it off the bottom and avoid this issue.

MEDITERRANEAN MOORING

Mediterranean mooring, also known as "med mooring" or "Tahitian mooring", is a technique for mooring a vessel to pier. In a Mediterranean mooring the vessel sets a temporary anchor off the pier and then approaches the pier at a perpendicular angle. The vessel then runs two lines to the pier. Alternatively, simple moorings may be placed off the pier and vessels may tie to these instead of setting a temporary anchor.

The advantage of Mediterranean mooring is that many more vessels can be connected to a fixed length of pier as they occupy only their width of pier rather than their length. The disadvantages of Mediterranean mooring are that it is more likely to result in collisions and that it is not practical in deep water or in regions with large tides.

MOORING LINE MATERIALS

Regular mooring lines

- Sisal
- Hemp
- Steel wire
- Polyethylene

- Polypropylene
- Polyester (e.g., used for deepsea mooring of offshore platforms)
- Nylon
- Chain

High-performance mooring lines

- HMPE (floating)
- Aramid (heat resistant) (including Kevlar)

ANCHOR

An anchor is a device normally made of metal, that is used to connect a vessel to the bed of a body of water to prevent the craft from drifting due to wind or current. The word derives from Latin ancora, which itself comes from the Greek (ankura).

Anchors can either be temporary or permanent. A permanent anchor is used in the creation of a mooring, and is rarely moved; a specialist service is normally needed to move or maintain it. Vessels carry one or more temporary anchors, which may be of different designs and weights. A sea anchor is a drogue, not in contact with the seabed, used to control a drifting vessel.

OVERVIEW

Anchors achieve holding power either by "hooking" into the seabed, or via sheer mass, or a combination of the two. Permanent moorings use large masses (commonly a block or slab of concrete) resting on the seabed. Semi-permanent mooring anchors (such as mushroom anchors) and large ship's anchors derive a significant portion of their holding power from their mass, while also hooking or embedding in the bottom. Modern anchors for smaller vessels have metal flukes which hook on to rocks on the bottom or bury themselves in soft seabed.

The vessel is attached to the anchor by the rode, which is made of chain, cable, rope, or a combination of these. The ratio of the length of rode to the water depth is known as the scope. Anchoring with sufficient scope and/or heavy chain rode brings the direction of strain close to parallel with the seabed. This is particularly important for light, modern anchors designed to bury in the bottom, where scopes of 5- to 7-to-1 are common, whereas heavy anchors and moorings can use a scope of 3-to-1, or less.

Since all anchors that embed themselves in the bottom require the strain to be along the seabed, anchors can be broken out of the bottom by shortening the rode until the vessel is directly above the anchor; at this point the anchor chain is "up and down", in naval parlance. If necessary, motoring slowly around the location of the anchor also helps dislodge it. Anchors are sometimes fitted with a tripping line attached to the crown, by which they can be unhooked from rocks or coral. The term aweigh describes an anchor when it is hanging on the rode and is not resting on the bottom. This is linked to the term to weigh anchor, meaning to lift the anchor from the sea bed, allowing the ship or boat to move. An anchor is described as aweigh when it has been broken out of the bottom and is being hauled up to be stowed. Aweigh should not be confused with under way, which describes a vessel which is not moored to a dock or anchored, whether or not the vessel is moving through the water.

EVOLUTION OF THE ANCHOR

The earliest anchors were probably rocks, and many rock anchors have been found dating from at least the Bronze Age. Many modern moorings still rely on a large rock as the primary element of their design. However, using pure mass to resist the forces of a storm only works well as a permanent mooring; a large enough rock would be nearly impossible to move to a new location.

The ancient Greeks used baskets of stones, large sacks filled with sand, and wooden logs filled with lead. According to Apollonius Rhodius and Stephen of Byzantium, anchors were formed of stone, and Athenaeus states that they were also sometimes made of wood. Such anchors held the vessel merely by their weight and by their friction along the bottom. Iron was afterwards introduced for the construction of anchors, and an improvement was made by forming them with teeth, or "flukes", to fasten themselves into the bottom.

ADMIRALTY PATTERN

The Admiralty Pattern, "A.P.", or simply "Admiralty", and also known as "Fisherman", is the anchor shape most familiar to non-sailors. It consists of a central shank with a ring or shackle for attaching the rode. At the other end of the shank there are two arms, carrying the flukes, while the stock is mounted to the other end, at ninety degrees to the arms. When the anchor lands on the bottom, it will generally fall over with the arms parallel to the seabed. As a strain comes onto the rode, the stock will dig into the bottom, canting the anchor until one of the flukes catches and digs into the bottom.

This basic design remained unchanged for centuries, with the most significant changes being to the overall proportions, and a move from stocks made of wood to iron stocks. Since one fluke always protrudes up from the set anchor, there is a great tendency of the rode to foul the anchor as the vessel swings due to wind or current shifts. When this happens, the anchor may be pulled out of the bottom, and in some cases may need to be hauled up to be re-set. In the mid-19th century, numerous modifications were attempted to alleviate these problems, as well as improve holding power, including one-armed mooring anchors. The most successful of these patent anchors, the Trotman Anchor, introduced a pivot where the arms join the shank, allowing the "idle" arm to fold against the shank. Handling and storage of these anchors requires special equipment and procedures. Once the anchor is hauled up to the hawsepipe, the ring end is hoisted up to the end of a timber projecting from the bow known as the cathead. The crown of the anchor is then hauled up with a heavy tackle until one fluke can be hooked over the rail. This is known as "catting and fishing" the anchor. Before dropping the anchor, the fishing process is reversed, and the anchor is dropped from the end of the cathead.

STOCKLESS ANCHOR

The stockless anchor, patented in England in 1821, represented the first significant departure in anchor design in centuries. Though their holding-power-to-weight ratio is significantly lower than admiralty pattern anchors, their ease of handling and stowage aboard large ships led to almost universal adoption.

In contrast to the elaborate stowage procedures for earlier anchors, stockless anchors are simply hauled up until they rest with the shank inside the hawsepipes, and the flukes against the hull (or inside a recess in the hull). While there are numerous variations, stockless anchors consist of a set of heavy flukes connected by a pivot or ball and socket joint to a shank. Cast into the crown of the anchor is a set of tripping palms, projections that drag on the bottom, forcing the main flukes to dig in.

SMALL BOAT ANCHORS

Until the mid-20th century, anchors for smaller vessels were either scaled-down versions of admiralty anchors, or simple grapnels. As new designs with greater holding-power-to-weight ratios, a great variety of anchor designs has emerged. Many of these designs are still under patent, and other types are best known by their original trademarked names.

GRAPNEL ANCHOR

A traditional design, the grapnel is merely a shank with four or more tines. It has a benefit in that, no matter how it reaches the bottom, one or more tines will be aimed to set. In coral it is often able to set quickly by hooking into the structure, but may be more difficult to retrieve. A grapnel is often quite light, and may have additional uses as a tool to recover gear lost overboard.

Its weight also makes it relatively easy to move and carry, however its shape is generally not very compact and it may be difficult to stow unless a collapsing model is used. Grapnels rarely have enough fluke area to develop much hold in sand, clay, or mud. It is not unknown for the anchor to foul on its own rode, or to foul the tines with refuse from the bottom, preventing it from digging in. On the other hand, it is quite possible for this anchor to find such a good hook that, without a trip line from the crown, it is impossible to retrieve.

HERRESHOFF ANCHOR

Designed by famous yacht designer L. Francis Herreshoff, this is essentially the same pattern as an admiralty anchor, albeit with small diamond shaped flukes or palms. The novelty of the design lay in the means by which it could be broken down into three pieces for stowage. In use, it still presents all the issues of the admiralty pattern anchor.

NORTHILL ANCHOR

Originally designed as a lightweight anchor for seaplanes, this design consists of two plow-like blades mounted to a shank, with a folding stock crossing through the crown of the anchor.

CQR (SECURE) PLOUGH ANCHOR

So named due to its resemblance to a traditional agricultural plough (or more specifically two ploughshares), many manufacturers produce a plough-style design, all based on or direct copies of the original CQR (Secure), a 1933 design patented in the UK (US patent in 1934) by mathematician Geoffrey Ingram Taylor. Ploughs are popular with cruising sailors and other private boaters. They are generally good in all bottoms, but not exceptional in any. The CQR design has a hinged shank, allowing the anchor to turn with direction changes rather than breaking out, while other plough types have a rigid shank. Plough anchors are usually stowed in a roller at the bow. Owing to the use of lead or other dedicated tip-weight, the plough is heavier than average for the amount of resistance developed, and may take a slightly longer pull to set thoroughly. It cannot be stored in a hawsepipe.

DELTA ANCHOR

The Delta was developed in the 1980s for commercialization by British marine manufacturer Simpson-Lawrence. It is similar in shape to the plough anchor but does not articulate.

DANFORTH OR FLUKE ANCHOR

American Richard Danforth invented the Danforth pattern in the 1940s for use aboard landing craft. It uses a stock at the crown to which two large flat triangular flukes are attached. The stock is hinged so the flukes can orient toward the bottom (and on some designs may be adjusted for an optimal angle depending on the bottom type). Tripping palms at the crown act to tip the flukes into the seabed. The design is a burying variety, and once well set can develop high resistance. Its light weight and compact flat design make it easy to retrieve and relatively easy to store; some anchor rollers and hawsepipes can accommodate a fluke-style anchor. The fluke anchor has difficulty penetrating kelp- and weed-covered bottoms, as well as rocky and particularly hard sand or clay bottoms. If there is much current, or the vessel is moving while dropping the anchor, it may "kite" or "skate" over the bottom due to the large fluke area acting as a sail or wing. Once set, the anchor tends to break out and reset when the direction of force changes dramatically, such as with the changing tide, and on some occasions it might not reset but instead drag.

BRUCE OR CLAW ANCHOR

This claw-shaped anchor was designed by Peter Bruce from the Isle of Man in the 1970s. Bruce gained his early reputation from the production of large-scale commercial anchors for ships and fixed installations such as oil rigs. The Bruce and its copies, known generically as "claws", have become a popular option for small boaters. It was intended to address some of the problems of the only general-purpose option then available, the plough. Claw-types set quickly in most seabeds and although not an articulated design, they have the reputation of not breaking out with tide or wind changes, instead slowly turning in the bottom to align with the force.

Claw types have difficulty penetrating weedy bottoms and grass. They offer a fairly low holding-power-to-weight ratio and generally have to be oversized to compete with other types. On the other hand they perform relatively well with low rode scopes and set fairly reliably. They cannot be used with hawsepipes.

RECENT DESIGNS

In recent years there has been something of a spurt in anchor design. Primarily designed to set very quickly, then generate high holding power, these anchors (mostly proprietary inventions still under patent) are finding homes with users of small to medium-sized vessels.

- The German-designed bow anchor, Bügelanker (or Wasi), has a sharp tip for penetrating weed, and features a roll-bar which allows the correct setting attitude to be achieved without the need for extra weight to be inserted into the tip.
- The Bulwagga is a unique design featuring three flukes instead of the usual two. It has performed well in tests by independent sources such as American boating magazine Practical Sailor.
- The Spade is a French design which has proved successful since 1996. It features a demountable shank and the choice of galvanized steel, stainless steel, or aluminium construction, which means a lighter and more easily stowable anchor.
- The New Zealand-designed Rocna has been produced since 2004. It too features a sharp toe like the Bügel for penetrating weed and grass, sets quickly, and has a particularly large fluke area. Its roll-bar is also similar to that of the Bügel. The Rocna obtained the highest averaged holding power in SAIL magazine's comparison

testing in 2006 and the highest "ultimate holding capacity" in Practical Boat Owner's testing in August 2011.

OTHER TEMPORARY ANCHORS

- Mud weight: Consists of a blunt heavy weight, usually cast iron or cast lead, that will sink into the mud and resist lateral movement. Suitable only for very soft silt bottoms and in mild conditions. Sizes range between 5 and 20 kg for small craft. Various designs exist and many are home produced from lead or improvised with heavy objects. This is a very commonly used method on the Norfolk Broads in England.

PERMANENT ANCHORS

These are used where the vessel is permanently or semi-permanently sited, for example in the case of lightvessels or channel marker buoys. The anchor needs to hold the vessel in all weathers, including the most severe storm, but needs to be lifted only occasionally, at most - for example, only if the vessel is to be towed into port for maintenance. An alternative to using an anchor under these circumstances, especially if the anchor need never be lifted at all, may be to use a pile driven into the seabed.

Permanent anchors come in a wide range of types and have no standard form. A slab of rock with an iron staple in it to attach a chain to would serve the purpose, as would any dense object of appropriate weight (for instance, an engine block). Modern moorings may be anchored by sand screws, which look and act very much like oversized screws drilled into the seabed, or by barbed metal beams pounded in (or even driven in with explosives) like pilings, or by a variety of other non-mass means of getting a grip on the bottom. One method of building a mooring is to use three or more conventional anchors laid out with short lengths of chain attached to a swivel, so no matter which direction the vessel moves, one or more anchors will be aligned to resist the force.

MUSHROOM ANCHOR

The mushroom anchor is suitable where the seabed is composed of silt or fine sand. It was invented by Robert Stevenson, for use by an 82-ton converted fishing boat, Pharos, which was used as a lightvessel between 1807 and 1810 near to Bell Rock whilst the lighthouse was being constructed. It was equipped with a 1.5-ton example.

It is shaped like an inverted mushroom, the head becoming buried in the silt. A counterweight is often provided at the other end of the shank to lay it down before it becomes buried. A mushroom anchor will normally sink in the silt to the point where it has displaced its own weight in bottom material, thus greatly increasing its holding power. These anchors are only suitable for a silt or mud bottom, since they rely upon suction and cohesion of the bottom material, which rocky or coarse sand bottoms lack. The holding power of this anchor is at best about twice its weight until it becomes buried, when it can be as much as ten times its weight. They are available in sizes from about 10 lb up to several tons.

DEADWEIGHT ANCHOR

This is an anchor which relies solely on being a heavy weight. It is usually just a large block of concrete or stone at the end of the chain. Its holding power is defined by its weight underwater (i.e. taking its buoyancy into account) regardless of the type of seabed, although suction can increase this if it becomes buried. Consequently deadweight anchors are used where mushroom anchors are unsuitable, for example in rock, gravel or coarse sand. An advantage of a deadweight anchor over a mushroom is that if it does become dragged, then it continues to provide its original holding force. The disadvantage of using deadweight anchors in conditions where a mushroom anchor could be used is that it needs to be around ten times the weight of the equivalent mushroom anchor.

SCREW ANCHOR

Screw anchors can be used to anchor permanent moorings, floating docks, fish farms, etc. These anchors must be screwed into the seabed with the use of a tool, so require access to the bottom, either at low tide or by use of a diver. Hence they can be difficult to install in deep water without special equipment. Weight for weight, screw anchors have a higher holding than other permanent designs, and so can be cheap and relatively easily installed, although may not be ideal in extremely soft mud.

ANCHORING GEAR

The elements of anchoring gear include the anchor, the cable (also called a rode), the method of attaching the two together, the method of attaching the cable to the ship, charts, and a method of learning the depth of the water. Vessels may carry a number of anchors: bower anchors (formerly known as sheet anchors) are the main anchors used by a vessel and normally carried at the bow of the vessel. A kedge anchor is a light anchor used for warping an anchor, also known as kedging, or more commonly on yachts for mooring quickly or in benign conditions.

A stream anchor, which is usually heavier than a kedge anchor, can be used for kedging or warping in addition to temporary mooring and restraining stern movement in tidal conditions or in waters where vessel movement needs to be restricted, such as rivers and channels. A Killick anchor is a small, possibly improvised, anchor. Charts are vital to good anchoring. Knowing the location of potential dangers, as well as being useful in estimating the effects of weather and tide in the anchorage, is essential in choosing a good place to drop the hook. One can get by without referring to charts, but they are an important tool and a part of good anchoring gear, and a skilled mariner would not choose to anchor without them.

The depth of water is necessary for determining scope, which is the ratio of length of cable to the depth measured from the highest point (usually the anchor roller or bow chock) to the seabed. For example, if the water is 25 ft (8 m) deep, and the anchor roller is 3 ft (1 m) above the water, the scope is the ratio between the amount of cable let out and 28 ft (9 m). For this reason it is important to have a reliable and accurate method of measuring the depth of water.

A cable or rode is the rope, chain, or combination thereof used to connect the anchor to the vessel. Chain rode is relatively heavy but resists abrasion. Fibre rope is more susceptible to abrasion on the seabed or obstructions, and is more likely to fail without warning. Combinations of a length of chain shackled to the anchor, with rope added to the other end of the chain are a common compromise on small craft.

ANCHORING TECHNIQUES

The basic anchoring consists of determining the location, dropping the anchor, laying out the scope, setting the hook, and assessing where the vessel ends up. The ship will seek a location which is sufficiently protected; has suitable holding ground, enough depth at low tide and enough room for the boat to swing.

The location to drop the anchor should be approached from down wind or down current, whichever is stronger. As the chosen spot is approached, the vessel should be stopped or even beginning to drift back. The anchor should be lowered quickly but under control until it is on the bottom. The vessel should continue to drift back, and the cable should be veered out under control so it will be relatively straight.

Once the desired scope is laid out, the vessel should be gently forced astern, usually using the auxiliary motor but possibly by backing a sail. A hand on the anchor line may telegraph a series of jerks and jolts, indicating the anchor is dragging, or a smooth tension indicative of digging in. As the anchor begins to dig in and resist backward force, the engine may be throttled up to get a thorough set. If the anchor continues to drag, or sets after having dragged too far, it should be retrieved and moved back to the desired position (or another location chosen.) There are techniques of anchoring to limit the swing of a vessel if the anchorage has limited room:

USING AN ANCHOR WEIGHT, KELLET OR SENTINEL

Lowering a concentrated, heavy weight down the anchor line - rope or chain - directly in front of the bow to the seabed, behaves like a heavy chain rode and lowers the angle of pull on the anchor. If the weight is suspended off the seabed it acts as a spring or shock absorber to dampen the sudden actions that are normally transmitted to the anchor and can cause it to dislodge and drag. In light conditions, a kellet will reduce the swing of the vessel considerably. In heavier conditions these effects disappear as the rode becomes straightened and the weight ineffective. Known as a "anchor chum weight" or "angel" in the UK.

FORKED MOOR

Using two anchors set approximately 45° apart, or wider angles up to 90°, from the bow is a strong mooring for facing into strong winds. To set anchors in this way, first one anchor is set in the normal fashion. Then, taking in on the first cable as the boat is motored into the wind and letting slack while drifting back, a second anchor is set approximately a half-scope away from the first on a line perpendicular to the wind.

After this second anchor is set, the scope on the first is taken up until the vessel is lying between the two anchors and the load is taken equally on each cable. This moor also to some degree limits the range of a vessel's swing to a narrower oval. Care should be taken that other vessels will not swing down on the boat due to the limited swing range.

BOW AND STERN

In the bow and stern technique, an anchor is set off each the bow and the stern, which can severely limit a vessel's swing range and also align it to steady wind, current or wave conditions.

One method of accomplishing this moor is to set a bow anchor normally, then drop back to the limit of the bow cable (or to double the desired scope, e.g. 8:1 if the eventual scope should be 4:1, 10:1 if the eventual scope should be 5:1, etc.) to lower a stern anchor. By taking up on the bow cable the stern anchor can be set. After both anchors are set, tension is taken up on both cables to limit the swing or to align the vessel.

BAHAMIAN MOOR

Similar to the above, a Bahamian moor is used to sharply limit the swing range of a vessel, but allows it to swing to a current. One of the primary characteristics of this technique is the use of a swivel as follows: the first anchor is set normally, and the vessel drops back to the limit of anchor cable. A second anchor is attached to the end of the anchor cable, and is dropped and set. A swivel is attached to the middle of the anchor cable, and the vessel connected to that.

The vessel will now swing in the middle of two anchors, which is acceptable in strong reversing currents, but a wind perpendicular to the current may break out the anchors, as they are not aligned for this load.

BACKING AN ANCHOR

Also known as tandem anchoring, in this technique two anchors are deployed in line with each other, on the same rode. With the foremost anchor reducing the load on the aft-most, this technique can develop great holding power and may be appropriate in "ultimate storm" circumstances. It does not limit swinging range, and might not be suitable in some circumstances. There are complications, and the technique requires careful preparation and a level of skill and experience above that required for a single anchor.

KEDGING

Kedging or warping is a technique for moving or turning a ship by using a relatively light anchor. In yachts, a kedge anchor is an anchor carried in addition to the main, or bower anchors, and usually stowed aft. Every yacht should carry at least two anchors - the main or bower anchor and a second lighter kedge anchor. It is used occasionally when it is necessary to limit the turning circle as the yacht swings when it is anchored, such as in a very narrow river or a deep pool in an otherwise shallow area.

For ships, a kedge may be dropped while a ship is underway, or carried out in a suitable direction by a tender or ship's boat to enable the ship to be winched off if aground or swung into a particular heading, or even to be held steady against a tidal or other stream. Historically, it was of particular relevance to sailing warships which used them to outmaneuver opponents when the wind had dropped but might be used by any vessel in confined, shoal water to place it in a more desirable position, provided she had enough manpower.

CLUB HAULING

Club hauling is an archaic technique. When a vessel is in a narrow channel or on a lee shore so that there is no room to tack the vessel in a conventional manner an anchor attached to the lee quarter may be dropped from the lee bow. This is deployed when the vessel is head to wind and has lost headway. As the vessel gathers sternway the strain on the cable pivots the vessel around what is now the weather quarter turning the vessel onto the other tack. The anchor is then normally cut away, as it cannot be recovered.

IN HERALDRY

An anchor frequently appears on the flags and coats of arms of institutions involved with the sea, both naval and commercial, as well as of port cities and seacoast regions and provinces in various countries. There also exists in heraldry the "Anchored Cross", or Mariner's Cross, a stylized cross in the shape of an anchor. The symbol can be used to signify 'fresh start' or 'hope'. The Mariner's Cross is also referred to as St. Clement's Cross, in reference to the way this saint was martyred (being tied to an anchor and thrown from a boat into the Black Sea in 102). Anchored crosses are occasionally a feature of coats of arms in which context they are referred to by the heraldic terms anchry or ancre.

MOORING MAST

A mooring mast, or mooring tower, is a structure designed to allow for the docking of an airship outside of an airship hangar or similar structure. More specifically, a mooring mast is a mast or tower that contains a fitting on its top that allows for the bow of the airship to attach its mooring line to the structure.

When it is not necessary or convenient to put an airship into its hangar (or shed) between flights, airships can be moored on the surface of land or water, in the air to one or more wires, or to a mooring mast. After their development mooring masts became the standard approach to mooring airships as considerable manhandling was avoided.

MAST TYPES

Airship mooring masts can be broadly divided into fixed high masts and fixed or mobile low (or 'stub') masts. In the 1920s and 1930s masts were built in many countries. At least two were mounted on ships. Without doubt the tallest mooring mast ever designed was the spire of the Empire State Building which was originally constructed to serve as a mooring mast, although soon after converted for use as a television and radio transmitter tower due to the discovered infeasibility of mooring an airship, for any length of time, to a very tall mast in the middle of an urban area.

EARLY DEVELOPMENTS

Mooring an airship by the nose to the top of a mast or tower of some kind might appear to be an obvious solution, but dirigibles had been flying for some years before the mooring mast made its appearance. The first airship known to have been moored to a mast was HMA (His Majesty's Airship) No.1, named the 'Mayfly', on 22 May 1911. The 38 ft (12 m) mast was mounted on a pontoon, and a windbreak of cross-yards with strips of canvas were attached to it. However, the windbreak caused the ship to yaw badly, and she became more stable when it was removed, withstanding winds gusting up to 43 miles per hour (69 km/h). Further experiments in mooring blimps to cable-stayed lattice masts were carried out during 1918.

BRITISH HIGH MAST OPERATIONS

The British mooring mast was developed more or less to its final form at Pulham and Cardington in 1919-21 to make the landing process easier and more economical of manpower The following account of the British high mast in its fully developed state at Cardington, and its operation, is abridged from

Masefield. Mooring masts were developed to act as a safe open harbour to which airships could be moored or unmoored in any weather, and at which they could receive (hydrogen or helium) gas, fuel, stores and payload.

The Cardington mast, completed in 1926, was an eight sided steel girder structure, 200 feet (61 m) high, tapering from 70 feet (21 m) diameter at ground level to 26 feet 6 inches (8.1 m) at the passenger platform, 170 feet (52 m) from the ground. Above the passenger platform was the 30 feet (9.1 m) of the conical housing for the mooring gear. A lower platform 142 feet (43 m) above the ground accommodated searchlights and signalling gear in a gallery 4 feet (1.2 m) wide. The top platform, at the height of 170 feet (52 m), from which passengers embarked and disembarked to and from the airships, was 40 feet (12 m) in diameter and encircled by a heavy parapet. The top rail of the parapet formed a track on which a gangway, let down from the airship, ran on wheels to give freedom for the airship to move around the tower as it swung with the wind. An electric passenger lift ran up the centre of the tower, encircled by a stairway to provide foot access.

The upper portion of the tower, from the passenger platform upwards, was a circular steel turret surmounted by a truncated cone with its top 23 feet (7.0 m) above the passenger platform. A three-part telescopic arm, mounted on gimbals, projected through an opening at the top, free to swing from the vertical in any direction up to 30 degrees of movement. The top of the arm consisted of a bell-shaped cup mounted to rotate on ball bearings. A cable extended through the bell-mouth which, linked to a cable dropped from airship to be moored, enabled the nose of the airship to be drawn down until a cone on the nose locked home into the cup and so secured the airship to the tower. The telescopic arm was then centred, locked in the vertical position, and made free to rotate on a vertical axis so the airship could swing, nose to tower, in any direction of the wind. In the machinery house at the base of the tower three steam-driven winches operated the hauling gear through drums 5 feet (1.5 m) in diameter to give cable hauling speeds of 50 feet a minute.

While an airship approached the mast slowly against the wind a mooring cable was let out from nose to the ground and linked, by a ground party, to the end of the mooring cable paid out from the mast head. The cable was then slowly wound in with the airship riding about 600 feet (180 m) above the mast and down wind, with one engine running astern to maintain a pull on the cable. At this point, two side wires - or 'yaw guys' - were also connected to cables taken from the nose of the airship to pulley blocks some hundreds of feet apart on the ground and thence to winches at the base of the mast.

All three cables were then wound in together, the main pull being taken on the mooring cable while the yaw guys steadied the ship. When all the cable had been wound in an articulated mooring cone on the nose of the airship locked home into the cup on the mast. The mast fitting was made free to rotate as the airship swung with the wind with freedom also for pitch and roll.

A gangway, like a drawbridge, which could be drawn up flush with the nose of the airship, was then let down with its free end resting on the parapet of the platform running round the mast. Passengers and crew boarded and disembarked from the ship under cover along this gangway. About twelve men were needed to moor an airship to a mast.

Four high masts of the Cardington type were built along the proposed British Empire Airship Service routes, at Cardington itself, at Montreal (Canada), Ismailia (Egypt) and Karachi (then India, now in Pakistan). None of these survive. Similar masts were proposed at sites in Australia, Ceylon (now Sri Lanka), Bombay, Keeling Islands, Kenya, Malta, at Ohakea in New Zealand, and in South Africa. The general site specifications can be found in documents produced by the British government.

GERMAN MAST TECHNIQUES

German mooring methods differed significantly from those adopted by the British. To quote Pugsley (1981): "the Germans, originally for ease of transport and for economy, developed a system using much lower masts. The nose of the ship was tethered as before to the mast head, which was only a little higher than the semi-diameter of the ship's hull. The lower fin at the stern was then fixed to a heavy carriage running on a circular railway track around the mast, and this carriage was powered so as to be able to move around the track to keep the ship head on to the wind. In the most sophisticated form, used by the Hindenburg, the rail system was linked to rails running from the mast straight into the airship shed, and the mast was powered so that the ship could be moved mechanically into the shed, complete with mast and stern carriage".

The following account of landing the German airship Graf Zeppelin is abridged from Dick and Robinson (1985): Before attempting a landing, contact was made by radio or flag signals with the ground crew to determine the ground temperature and wind conditions. For a normal calm weather landing the ship was trimmed very slightly nose down, as this gave a better gliding angle and the ship almost flew herself down. A smoky fire was started on the ground to show the wind direction. The ship then made a long approach with a rate of fall of 100 feet per minute, and the lines were dropped when she was over the landing flag. When conditions were unusual, as in gusty and bumpy weather, the Graf was weighed off a little light, and the approach had to be fast and preferably long and low.

When the airship was over the field the engines had to be reversed for some time to stop her, and this also helped to get the nose down. Yaw lines dropped from the ship's nose were drawn out to Port and Starboard by thirty men each, while twenty more on each side pulled the ship down with spider lines (so called because twenty short lines radiated like the legs of a spider from a block). When the airship reached the ground, fifty men held the control car rails and twenty held those of the after car. With thirty men in reserve, the ground crew totalled two hundred men.

The ground crew would then walk the Graf to a short, or 'stub', mast, to which the nose of the airship would be attached. The airship would then rest on the ground with its rear gondola attached to a movable weighted carriage that enabled the airship to swing around the mast with the wind. In some places the stub mast was mounted on rails and could be drawn into the airship hangar, guiding the nose of the ship while the tail was controlled by the carriage attached to the rear gondola. Airships designed for landing on the ground had pneumatic bumper bags or undercarriage wheels under the main and rear gondolas (or tail fin).

Dick states that the Germans never used mooring masts until the Graf Zeppelin entered service in 1928, and never moored to high masts. To some extent this probably reflects the conservatism of the Zeppelin company operations. Long experience in handling airships in all sorts of conditions was valued and innovations or significant changes in practice were unlikely to be adopted unless clear advantages were apparent.

UNITED STATES

In the US a mix of techniques were applied, and airships moored to both high and stub masts. Large ground crews (or 'landing parties') of up to 340 men were required to manage the large airships Akron and Macon at landing or on the ground, before they could be attached to the stub mast. Being part of a ground crew was not risk-free. In gusty conditions, or if mis-handled, an airship could suddenly rise. If the ground crew did not immediately let go of the handling lines they risked being carried off their feet. In one famous incident captured on movie film in 1932, during the landing of the US airship Akron, three men were carried off their feet in this way, two to fall to their deaths after a short time. The third managed to improve his hold on the handling rope until he could be hauled into the airship.

SHIP-MOUNTED MOORING MASTS

At least two ships have mounted mooring masts. As the US intended to use large airships for long-range maritime patrol operations experiments were made in mooring airships to a mast mounted on the ship USS Patoka. Over time the airships USS Shenandoah, Los Angeles, and Akron all moored to the mast mounted at the stern of the ship, and operated using her as a base for resupply, refuelling and gassing.

The Spanish seaplane carrier Dédalo (1922-1935) carried a mooring mast at the bow to cater for small dirigibles carried on board. Around 1925 the Royal Navy considered the monitor HMS Roberts for conversion to a mobile airship base with a mooring mast and fueling capabilities, but nothing came of this proposal.

MODERN MAST OPERATIONS

Smaller mobile masts have been used for small airships and blimps for a long time. They may be wheel or track-mounted, and can be operated by a small crew. The general operating principle is broadly similar to the larger masts. Modern blimps may operate from mobile masts for months at a time without returning to their hangars.

6

Wind Wave

In fluid dynamics, wind waves or, more precisely, wind-generated waves are surface waves that occur on the free surface of oceans, seas, lakes, rivers, and canals or even on small puddles and ponds. They usually result from the wind blowing over a vast enough stretch of fluid surface. Waves in the oceans can travel thousands of miles before reaching land. Wind waves range in size from small ripples to huge waves over 30 m high. When directly generated and affected by local winds, a wind wave system is called a wind sea. After the wind ceases to blow, wind waves are called swells. More generally, a swell consists of wind-generated waves that are not-or are hardly-affected by the local wind at that time. They have been generated elsewhere or some time ago. Wind waves in the ocean are called ocean surface waves.

Wind waves have a certain amount of randomness: subsequent waves differ in height, duration, and shape with limited predictability. They can be described as a stochastic process, in combination with the physics governing their generation, growth, propagation and decay-as well as governing the interdependence between flow quantities such as: the water surface movements, flow velocities and water pressure. The key statistics of wind waves (both seas and swells) in evolving sea states can be predicted with wind wave models. Tsunamis are a specific type of wave not caused by wind but by geological effects. In deep water, tsunamis are not visible because they are small in height and very long in wavelength. They may grow to devastating proportions at the coast due to reduced water depth.

WAVE FORMATION

The great majority of large breakers one observes on a beach result from distant winds. Five factors influence the formation of wind waves:

- Wind speed
- Distance of open water that the wind has blown over (called the fetch)
- Width of area affected by fetch
- Time duration the wind has blown over a given area
- Water depth

All of these factors work together to determine the size of wind waves. The greater each of the variables, the larger the waves. Waves are characterized by:

- Wave height (from trough to crest)
- Wave length (from crest to crest)
- Wave period (time interval between arrival of consecutive crests at a stationary point)
- Wave propagation direction

Waves in a given area typically have a range of heights. For weather reporting and for scientific analysis of wind wave statistics, their characteristic height over a period of time is usually expressed as significant wave height. The significant wave height is also the value a "trained observer" (e.g. from a ship's crew) would estimate from visual observation of a sea state. Given the variability of wave height, the largest individual waves are likely to be somewhat less than twice the reported significant wave height for a particular day or storm.

TYPES OF WIND WAVES

Three different types of wind waves develop over time:

- Capillary waves, or ripples
- Seas
- Swells

Ripples appear on smooth water when the wind blows, but will die quickly if the wind stops. The restoring force that allows them to propagate is surface tension. Seas are the larger-scale, often irregular motions that form under sustained winds. These waves tend to last much longer, even after the wind has died, and the restoring force that allows them to propagate is gravity. As waves propagate away from their area of origin, they naturally separate into groups of common direction and wavelength. The sets of waves formed in this way are known as swells.

Individual "rogue waves" (also called "freak waves", "monster waves", "killer waves", and "king waves") much higher than the other waves in the sea state can occur. In the case of the Draupner wave, its 25 m (82 ft) height was 2.2 times the significant wave height. Such waves are distinct from tides, caused by the Moon and Sun's gravitational pull, tsunamis that are caused by underwater earthquakes or landslides, and waves generated by underwater explosions or the fall of meteorites-all having far longer wavelengths than wind waves. Yet, the largest ever recorded wind waves are common-not rogue-waves in extreme sea states. For example: 29.1 m (95 ft) high waves have been recorded on the RRS Discovery in a sea with 18.5 m (61 ft) significant wave height, so the highest wave is only 1.6 times the significant wave height. The biggest recorded by a buoy (as of 2011) was 32.3 m (106 ft) high during the 2007 typhoon Krosa near Taiwan.

WIND WAVE MODELS

Surfers are very interested in the wave forecasts. There are many websites that provide predictions of the surf quality for the upcoming days and weeks. Wind wave models are driven by more general weather models that predict the winds and pressures over the oceans, seas and lakes.

Wind wave models are also an important part of examining the impact of shore protection and beach nourishment proposals. For many beach areas there is only patchy information about the wave climate, therefore estimating the effect of wind waves is important for managing littoral environments.

WAVE SHOALING

In fluid dynamics, wave shoaling is the effect by which surface waves entering shallower water increase in wave height (which is about twice the amplitude). It is caused by the fact that the group velocity, which is also the wave-energy transport velocity, decreases with the reduction of water depth. Under stationary conditions, this decrease in transport speed must be compensated by an increase in energy density in order to maintain a constant energy flux. Shoaling waves will also exhibit a reduction in wavelength while the frequency remains constant. In shallow water and parallel depth contours, non-breaking waves will increase in wave height as the wave packet enters shallower water. This is particularly evident for tsunamis as they wax in height when approaching a coastline, with devastating results.

REFRACTION

Refraction is the change in direction of a wave due to a change in its transmission medium. Refraction is essentially a surface phenomenon. The phenomenon is mainly in governance to the law of conservation of energy and momentum. Due to change of medium, the phase velocity of the wave is changed but its frequency remains constant. This is most commonly observed when a wave passes from one medium to another at any angle other than 90° or 0°. Refraction of light is the most commonly observed phenomenon, but any type of wave can refract when it interacts with a medium, for example when sound waves pass from one medium into another or when water waves move into water of a different depth. Refraction is described by Snell's law, which states that for a given pair of media and a wave with a single frequency, the ratio of the sines of the angle of incidence ?1 and angle of refraction ?2 is equivalent to the ratio of phase velocities (v1 / v2) in the two media, or equivalently, to the opposite ratio of the indices of refraction (n2 / n1):

$$\frac{\sin t_1}{\sin t_2} @ \frac{v_1}{v_2} @ \frac{n_2}{n_1}.$$

In general, the incident wave is partially refracted and partially reflected; the details of this behavior are described by the Fresnel equations.

EXPLANATION

In optics, refraction is a phenomenon that often occurs when waves travel from a medium with a given refractive index to a medium with another at an oblique angle. At the boundary between the media, the wave's phase velocity is altered, usually causing a change in direction. Its wavelength increases or decreases but its frequency remains constant. For example, a light ray will refract as it enters and leaves glass, assuming there is a change in refractive index. A ray traveling along the normal (perpendicular to the boundary) will change speed, but not direction. Refraction still occurs in this case. Understanding of this concept led to the invention of lenses and the refracting telescope.

Refraction can be seen when looking into a bowl of water. Air has a refractive index of about 1.0003, and water has a refractive index of about 1.3330. If a person looks at a straight object, such as a pencil or straw, which is placed at a slant, partially in the water, the object appears to bend at the water's surface. This is due to the bending of light rays as they move from the water to the air. Once the rays reach the eye, the eye traces them back as straight lines (lines of sight). The lines of sight (shown as dashed lines) intersect at a higher position than where the actual rays originated. This causes the pencil to appear higher and the water to appear shallower than it really is. The depth that the water appears to be when viewed from above is known as the apparent depth.

This is an important consideration for spearfishing from the surface because it will make the target fish appear to be in a different place, and the fisher must aim lower to catch the fish. Conversely, an object above the water has a higher apparent height when viewed from below the water. The opposite correction must be made by an archer fish. For small angles of incidence (measured from the normal, when sin t is approximately the same as tan t), the ratio of apparent to real depth is the ratio of the refractive indexes of air to that of water. But as the angle of incidence approaches 90o, the apparent depth approaches zero, albeit reflection increases, which limits observation at high angles of incidence. Conversely, the apparent height approaches infinity as the angle of incidence (from below) increases, but even earlier, as the angle of total internal reflection is approached, albeit the image also fades from view as this limit is approached.

The diagram on the right shows an example of refraction in water waves. Ripples travel from the left and pass over a shallower region inclined at an angle to the wavefront. The waves travel slower in the more shallow water, so the wavelength decreases and the wave bends at the boundary. The dotted line represents the normal to the boundary. The dashed line represents the original direction of the waves. This phenomenon explains why waves on a shoreline tend to strike the shore close to a perpendicular angle. As the waves travel from deep water into shallower water near the shore, they are refracted from their original direction of travel to an angle more normal to the shoreline. Refraction is also responsible for rainbows and for the splitting of white light into a rainbow-spectrum as it passes through a glass prism. Glass has a higher refractive index than air. When a beam of white light passes from air into a material having an index of refraction that varies with frequency, a phenomenon known as dispersion occurs, in which different coloured components of the white light are refracted at different angles, i.e., they bend by different amounts at the interface, so that they become separated. The different colors correspond to different frequencies.

While refraction allows for phenomena such as rainbows, it may also produce peculiar optical phenomena, such as mirages and Fata Morgana. These are caused by the change of the refractive index of air with temperature. The refractive index of materials can also be nonlinear, as occurs with the Kerr effect when high intensity light leads to a refractive index proportional to the intensity of the incident light.

Recently some metamaterials have been created which have a negative refractive index. With metamaterials, we can also obtain total refraction phenomena when the wave impedances of the two media are matched. There is then no reflected wave. Also, since refraction can make objects appear closer than they are, it is responsible for allowing water to magnify objects. First, as light is entering a drop of water, it slows down. If the water's surface is not flat, then the light will be bent into a new path. This round shape will bend the light outwards and as it spreads out, the image you see gets larger.

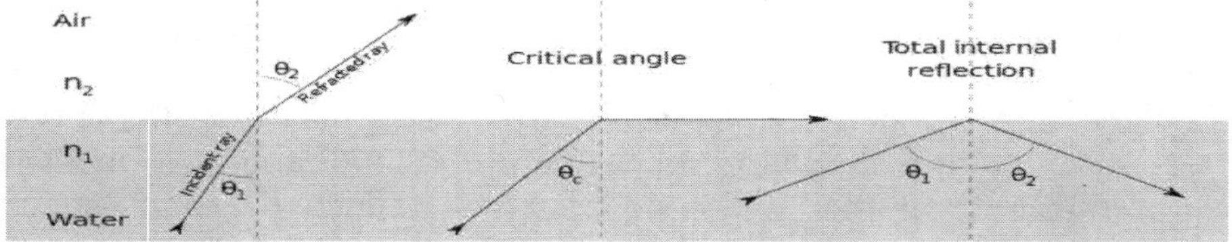

Fig. Refraction of light at the interface between two media.

A useful analogy in explaining the refraction of light would be to imagine a marching band as they march at an oblique angle from pavement (a fast medium) into mud (a slower medium). The marchers on the side that runs into the mud first will slow down first. This causes the whole band to pivot slightly toward the normal (make a smaller angle from the normal).

CLINICAL SIGNIFICANCE

In medicine, particularly optometry, ophthalmology and orthoptics, refraction (also known as refractometry) is a clinical test in which a phoropter may be used by the appropriate eye care professional to determine the eye's refractive error and the best corrective lenses to be prescribed. A series of test lenses in graded optical powers or focal lengths are presented to determine which provides the sharpest, clearest vision.

ACOUSTICS

In underwater acoustics, refraction is the bending or curving of a sound ray that results when the ray passes through a sound speed gradient from a region of one sound speed to a region of a different speed. The amount of ray bending is dependent upon the amount of difference between sound speeds, that is, the variation in temperature, salinity, and pressure of the water. Similar acoustics effects are also found in the Earth's atmosphere. The phenomenon of refraction of sound in the atmosphere has been known for centuries; however, beginning in the early 1970s, widespread analysis of this effect came into vogue through the designing of urban highways and noise barriers to address the meteorological effects of bending of sound rays in the lower atmosphere.

AIRY WAVE THEORY

In fluid dynamics, Airy wave theory (often referred to as linear wave theory) gives a linearised description of the propagation of gravity waves on the surface of a homogeneous fluid layer. The theory assumes that the fluid layer has a uniform mean depth, and that the fluid flow is inviscid, incompressible and irrotational. This theory was first published, in correct form, by George Biddell Airy in the 19th century. Airy wave theory is often applied in ocean engineering and coastal engineering for the modelling of random sea states - giving a description of the wave kinematics and dynamics of high-enough accuracy for many purposes. Further, several second-order nonlinear properties of surface gravity waves, and their propagation, can be estimated from its results. Airy wave theory is also a good approximation for tsunami waves in the ocean, before they steepen near the coast.

This linear theory is often used to get a quick and rough estimate of wave characteristics and their effects. This approximation is accurate for small ratios of the wave height to water depth (for waves in shallow water), and wave height to wavelength (for waves in deep water).

DESCRIPTION

Airy wave theory uses a potential flow (or velocity potential) approach to describe the motion of gravity waves on a fluid surface. The use of - inviscid and irrotational - potential flow in water waves is remarkably successful, given its failure to describe many other fluid flows where it is often essential to take viscosity, vorticity, turbulence and/or flow separation into account. This is due to the fact that for the oscillatory part of the fluid motion, wave-induced vorticity is restricted to some thin oscillatory Stokes boundary layers at the boundaries of the fluid domain.

Airy wave theory is often used in ocean engineering and coastal engineering. Especially for random waves, sometimes called wave turbulence, the evolution of the wave statistics - including the wave spectrum - is predicted well over not too long distances (in terms of wavelengths) and in not too shallow water. Diffraction is one of the wave effects which can be described with Airy wave theory. Further, by using the WKBJ approximation, wave shoaling and refraction can be predicted.

Earlier attempts to describe surface gravity waves using potential flow were made by, among others, Laplace, Poisson, Cauchy and Kelland. But Airy was the first to publish the correct derivation and formulation in 1841. Soon after, in 1847, the linear theory of Airy was extended by Stokes for non-linear wave motion - known as Stokes' wave theory - correct up to third order in the wave steepness. Even before Airy's linear theory, Gerstner derived a nonlinear trochoidal wave theory in 1804, which however is not irrotational.

Airy wave theory is a linear theory for the propagation of waves on the surface of a potential flow and above a horizontal bottom. The free surface elevation k(x,t) of one wave component is sinusoidal, as a function of horizontal position x and time t:

$$k(x,t)@a\cos\left(kx\,0\,zt\right)$$

where

- a is the wave amplitude in metre,
- cos is the cosine function,
- k is the angular wavenumber in radian per metre, related to the wavelength o as

$$k@\frac{2s}{o},$$

- z is the angular frequency in radian per second, related to the period T and frequency f by

$$z@\frac{2s}{T}@2s\,f.$$

The waves propagate along the water surface with the phase speed c_p:

$$c_p@\frac{z}{k}@\frac{c}{T}.$$

The angular wavenumber k and frequency z are not independent parameters (and thus also wavelength o and period T are not independent), but are coupled. Surface gravity waves on a fluid are dispersive waves - exhibiting frequency dispersion - meaning that each wavenumber has its own frequency and phase speed.

Note that in engineering the wave height H - the difference in elevation between crest and trough - is often used:

$$H@2a \qquad \text{and} \qquad a@\frac{1}{2}H,$$

valid in the present case of linear periodic waves.

Underneath the surface, there is a fluid motion associated with the free surface motion. While the surface elevation shows a propagating wave, the fluid particles are in an orbital motion. Within the framework of Airy wave theory, the orbits are closed curves: circles in deep water, and ellipses in finite depth-with the ellipses becoming flatter near the bottom of the fluid layer. So while the wave propagates, the fluid particles just orbit (oscillate) around their average position. With the propagating wave motion, the fluid particles transfer energy in the wave propagation direction, without having a mean velocity. The diameter of the orbits reduces with depth below the free surface. In deep water, the orbit's diameter is reduced to 4% of its free-surface value at a depth of half a wavelength.

In a similar fashion, there is also a pressure oscillation underneath the free surface, with wave-induced pressure oscillations reducing with depth below the free surface - in the same way as for the orbital motion of fluid parcels.

MATHEMATICAL FORMULATION OF THE WAVE MOTION

Flow problem formulation

The waves propagate in the horizontal direction, with coordinate x, and a fluid domain bound above by a free surface at z = k(x,t), with z the vertical coordinate (positive in the upward direction) and t being time. The level z = 0 corresponds with the mean surface elevation.

The impermeable bed underneath the fluid layer is at z = -h. Further, the flow is assumed to be incompressible and irrotational - a good approximation of the flow in the fluid interior for waves on a liquid surface - and potential theory can be used to describe the flow. The velocity potential ϕ (x,z,t) is related to the flow velocity components u_x and u_z in the horizontal (x) and vertical (z) directions by:

$$u_x = \frac{\partial \phi}{\partial x} \quad \text{and} \quad u_z = \frac{\partial \phi}{\partial z}.$$

Then, due to the continuity equation for an incompressible flow, the potential ϕ has to satisfy the Laplace equation:

$$\frac{\partial^2 \phi}{\partial x^2} + \frac{\partial^2 \phi}{\partial z^2} = 0. \quad (1)$$

Boundary conditions are needed at the bed and the free surface in order to close the system of equations. For their formulation within the framework of linear theory, it is necessary to specify what the base state (or zeroth-order solution) of the flow is. Here, we assume the base state is rest, implying the mean flow velocities are zero. The bed being impermeable, leads to the kinematic bed boundary-condition:

$$\frac{\partial \Phi}{\partial z} = 0 \quad \text{at } z = -h. \tag{2}$$

In case of deep water - by which is meant infinite water depth, from a mathematical point of view - the flow velocities have to go to zero in the limit as the vertical coordinate goes to minus infinity: $z \to -\infty$.

At the free surface, for infinitesimal waves, the vertical motion of the flow has to be equal to the vertical velocity of the free surface. This leads to the kinematic free-surface boundary-condition:

$$\frac{\partial \eta}{\partial t} = \frac{\partial \Phi}{\partial z} \quad \text{at } z = \eta(x,t). \tag{3}$$

If the free surface elevation $\eta(x,t)$ was a known function, this would be enough to solve the flow problem. However, the surface elevation is an extra unknown, for which an additional boundary condition is needed. This is provided by Bernoulli's equation for an unsteady potential flow. The pressure above the free surface is assumed to be constant. This constant pressure is taken equal to zero, without loss of generality, since the level of such a constant pressure does not alter the flow. After linearisation, this gives the dynamic free-surface boundary condition:

$$\frac{\partial \Phi}{\partial t} + g\eta = 0 \quad \text{at } z = \eta(x,t). \tag{4}$$

Because this is a linear theory, in both free-surface boundary conditions - the kinematic and the dynamic one, equations (3) and (4) - the value of Φ and $\partial \Phi / \partial z$ at the fixed mean level z = 0 is used.

TABLE OF WAVE QUANTITIES

In the table below, several flow quantities and parameters according to Airy wave theory are given. The given quantities are for a bit more general situation as for the solution given above. Firstly, the waves may propagate in an arbitrary horizontal direction in the x = (x,y) plane. The wavenumber vector is k, and is perpendicular to the cams of the wave crests. Secondly, allowance is made for a mean flow velocity U, in the horizontal direction and uniform over (independent of) depth z.

This introduces a Doppler shift in the dispersion relations. At an Earth-fixed location, the observed angular frequency (or absolute angular frequency) is ω. On the other hand, in a frame of reference moving with the mean velocity U (so the mean velocity as observed from this reference frame is zero), the angular frequency is different. It is called the intrinsic angular frequency (or relative angular frequency), denoted as σ. So in pure wave motion, with U=0, both frequencies ω and σ are equal. The wave number k (and wavelength λ) are independent of the frame of reference, and have no Doppler shift (for monochromatic waves). The table only gives the oscillatory parts of flow quantities - velocities, particle excursions and pressure - and not their mean value or drift. The oscillatory particle excursions $\{_x$ and $\{_z$ are the time integrals of the oscillatory flow velocities ux and uz respectively.

Water depth is classified into three regimes:

- deep water - for a water depth larger than half the wavelength, h>½o, the phase speed of the waves is hardly influenced by depth (this is the case for most wind waves on the sea and ocean surface),
- shallow water - for a water depth smaller than the wavelength divided by 20, h<½20 o, the phase speed of the waves is only dependent on water depth, and no longer a function of period or wavelength; and
- intermediate depth - all other cases, $/_{20}$ o<h < ½ o, where both water depth and period (or wavelength) have a significant influence on the solution of Airy wave theory.

In the limiting cases of deep and shallow water, simplifying approximations to the solution can be made. While for intermediate depth, the full formulations have to be used.

TWO LAYERS BOUNDED ABOVE BY A FREE SURFACE

In this case the dispersion relation allows for two modes: a barotropic mode where the free surface amplitude is large compared with the amplitude of the interfacial wave, and a baroclinic mode where the opposite is the case - the interfacial wave is higher than and in antiphase with the free surface wave. The dispersion relation for this case is of a more complicated form.

SECOND-ORDER WAVE PROPERTIES

Several second-order wave properties, i.e. quadratic in the wave amplitude a, can be derived directly from Airy wave theory. They are of importance in many practical applications, e.g. forecasts of wave conditions. Using a WKBJ approximation, second-order wave properties also find their applications in describing waves in case of slowly varying bathymetry, and mean-flow variations of currents and surface elevation. As well as in the description of the wave and mean-flow interactions due to time and space-variations in amplitude, frequency, wavelength and direction of the wave field itself.

BREAKING WAVE

In fluid dynamics, a breaking wave is a wave whose amplitude reaches a critical level at which some process can suddenly start to occur that causes large amounts of wave energy to be transformed into turbulent kinetic energy. At this point, simple physical models that describe wave dynamics often become invalid, particularly those that assume linear behaviour. The most generally familiar sort of breaking wave is the breaking of water surface waves on a coastline. Because of the horizontal component of the fluid velocity associated with the wave motion, wave crests steepen as the amplitude increases; wave breaking generally occurs where the amplitude reaches the point that the crest of the wave actually overturns-though the types of breaking water surface waves are discussed in more detail below. Certain other effects in fluid dynamics have also been termed "breaking waves," partly by analogy with water surface waves. In meteorology, atmospheric gravity waves are said to break when the wave produces regions where the potential temperature decreases with height, leading to energy dissipation through convective instability; likewise Rossby waves are said to break when the potential vorticity gradient is overturned. Wave breaking also occurs in plasmas, when the particle velocities exceed the wave's phase speed.

BREAKING WATER SURFACE WAVES

Breaking of water surface waves may occur anywhere that the amplitude is sufficient, including in mid-ocean. However, it is particularly common on beaches because wave heights are amplified in the region of shallower water (because the group velocity is lower there). See also waves and shallow water.

There are four basic types of breaking water waves. They are spilling, plunging, collapsing, and surging.

Spilling breakers

When the ocean floor has a gradual slope, the wave will steepen until the crest becomes unstable, resulting in turbulent whitewater spilling down the face of the wave. This continues as the wave approaches the shore, and the wave's energy is slowly dissipated in the whitewater. Because of this, spilling waves break for a longer time than other waves, and create a relatively gentle wave. Onshore wind conditions make spillers more likely.

Plunging breakers

A plunging wave occurs when the ocean floor is steep or has sudden depth changes, such as from a reef or sandbar. The crest of the wave becomes much steeper than a spilling wave, becomes vertical, then curls over and drops onto the trough of the wave, releasing most of its energy at once in a relatively violent impact. A plunging wave breaks with more energy than a significantly larger spilling wave. The wave can trap and compress the air under the lip, which creates the "crashing" sound associated with waves. With large waves, this crash can be felt by beachgoers on land. Offshore wind conditions can make plungers more likely.

If a plunging wave is not parallel to the beach (or the ocean floor), the section of the wave which reaches shallow water will break first, and the breaking section (or curl) will move laterally across the face of the wave as the wave continues. This is the "tube" that is so highly sought after by surfers (also called a "barrel", a "pit", and "the greenroom", among other terms). The surfer tries to stay near or under the curling lip, often trying to stay as "deep" in the tube as possible while still being able to shoot forward and exit the tube before it closes. A plunging wave that is parallel to the beach can break along its whole length at once, rendering it unrideable and dangerous. Surfers refer to these waves as "closed out".

Collapsing

Collapsing waves are a cross between plunging and surging, in which the crest never fully breaks, yet the bottom face of the wave gets steeper and collapses, resulting in foam.

Surging

Surging breakers originate from long period, low steepness waves and/ or steep beach profiles. The outcome is the rapid movement of the base of the wave up the swash slope and the disappearance of the wave crest. The front face and crest of the wave remain relatively smooth with little foam or bubbles, resulting in a very narrow surf zone, or no breaking waves at all. The short, sharp burst of wave energy means that the swash/backwash cycle completes before the arrival of the next wave, leading to a low value of Kemp's phase difference (< 0.5). Surging waves are typical of reflective beach states. On steeper beaches, the energy of the wave can be reflected by the bottom back into the ocean, causing standing waves.

PHYSICS

During breaking, a deformation (usually a bulge) forms at the wave crest, either leading side of which is known as the "toe." Parasitic capillary waves are formed, with short wavelengths. Those above the "toe" tend to have much longer wavelengths. This theory is anything but perfect, however, as it's linear. There have been a couple non-linear theories of motion (regarding waves). One put forth uses a perturbation method to expand the description all the way to the third order, and better solutions have been found since then. As for wave deformation, methods much like the boundary integral method and the Boussinesq model have been created.

It has been found that high-frequency detail present in a breaking wave plays a part in crest deformation and destabilization. The same theory expands on this, stating that the valleys of the capillary waves create a source for vorticity. It is said that surface tension (and viscosity) are significant for waves up to 2m in wavelength.

These models are flawed, however, as they can't take into account what happens to the water after the wave breaks. Post-break eddy forms and the turbulence created via the breaking is mostly unresearched. Understandably, it might be difficult to glean predictable results from the ocean. After the tip of the wave overturns and the jet collapses, it creates a very coherent and defined horizontal vertex. The plunging breakers create secondary eddies down the face of the wave. Small horizontal random eddides that form on the sides of the wave suggest that, perhaps, prior to breaking, the water's velocity is more or less two dimensional. This becomes three dimensional upon breaking.

The main vortex along the front of the wave diffuses rapidly into the interior of the wave after breaking, as the eddies on the surface become more viscous. Advection and molecular diffusion play a part in stretching the vortex and redistributing the vorticity, as well as the formation turbulence cascades. The energy of the large vortices are, by this method, is transferred to much smaller isotropic vortices. Experiments have been conducted to deduce the evolution of turbulence after break, both in deep water and on a beach.

WIND WAVE MODEL

In fluid dynamics, wind wave modeling describes the effort to depict the sea state and predict the evolution of the energy of wind waves using numerical techniques. These simulations consider atmospheric wind forcing, nonlinear wave interactions, and frictional dissipation, and they output statistics describing wave heights, periods, and propagation directions for regional seas or global oceans. Such wave hindcasts and wave forecasts are extremely important for commercial interests on the high seas. For example, the shipping industry requires guidance for operational planning and tactical seakeeping purposes.

For the specific case of predicting wind wave statistics on the ocean, the term ocean surface wave model is used. Other applications, in particular coastal engineering, have led to the developments of wind wave models specifically designed for coastal applications.

HISTORICAL OVERVIEW

Early forecasts of the sea state were created manually based upon empirical relationships between the present state of the sea, the expected wind conditions, the fetch/duration, and the direction of the wave propagation. Alternatively, the swell part of the state has been forecasted as early as 1920 using remote observations.

During the 1950s and 1960s, much of the theoretical groundwork necessary for numerical descriptions of wave evolution was laid. For forecasting purposes, it was realized that the random nature of the sea state was best described by a spectral decomposition in which the energy of the waves was attributed to as many wave trains as necessary, each with a specific direction and period. This approach allowed to make combined forecasts of wind seas and swells. The first numerical model based on the spectral decomposition of the sea state was operated in 1956 by the French Weather Service, and focused on the North Atlantic. The 1970s saw the first operational, hemispheric wave model: the spectral wave ocean model (SWOM) at the Fleet Numerical Oceanography Center. First generation wave models did not consider nonlinear wave interactions. Second generation models, available by the early 1980s, parameterized these interactions. They included the "coupled hybrid" and "coupled discrete" formulations. Third generation models explicitly represent all the physics relevant for the development of the sea state in two dimensions. The wave modeling project (WAM), an international effort, led to the refinement of modern wave modeling techniques during the decade 1984-1994. Improvements included two-way coupling between wind and waves, assimilation of satellite wave data, and medium-range operational forecasting.

Wind wave models are used in the context of a forecasting or hindcasting system. Differences in model results arise, with decreasing order of importance, from differences in wind and sea ice forcing, differences in parameterizations of physical processes, the use of data assimilation and associated methods, the numerical techniques used to solve the wave energy evolution equation.

GENERAL STRATEGY

Input

A wave model requires as initial conditions information describing the state of the sea. An analysis of the sea or ocean can be created through data assimilation, where observations such as buoy or satellite altimeter measurements are combined with a background guess from a previous forecast or climatology to create the best estimate of the current conditions. In practice, many forecasting system rely only on the previous forecast, without any assimilation of observations.

A more critical input is the "forcing" by wind fields: a time-varying map of wind speed and directions. The most common sources of errors in wave model results are the errors in the wind field. Ocean currents can also be important, in particular in western boundary currents such as the Gulf Stream, Kuroshio or Agulhas current, or in coastal areas where tidal currents are strong. Waves are also affected by sea ice and icebergs, and all operational global wave models take at least the sea ice into account.

Representation

The sea state is described as a spectrum; the sea surface can be decomposed into waves of varying frequencies using the principle of superposition. The waves are also separated by their direction of propagation. The model domain size can range from regional to the global ocean. Smaller domains can be nested within a global domain to provide higher resolution in a region of interest. The sea state evolves according to physical equations, which include wave propagation / advection, and a source function which allows for wave energy to be augmented or diminished. The source function has at least three terms: wind forcing, nonlinear transfer, and dissipation by whitecapping. Wind data are typically provided from a separate atmospheric model from an operational weather forecasting center.

For intermediate water depths the effect of bottom friction should also be added. At ocean scales, the dissipation of swells - without breaking- is a very important term.

Output

The output of a wind wave model is a description of the wave spectra, with amplitudes associated with each frequency and propagation direction. Results are typically summarized by the significant wave height, which is the average height of the one-third largest waves, and the period and propagation direction of the dominant wave.

Coupled models

Wind waves also act to modify atmospheric properties through frictional drag of near-surface winds and heat fluxes. Two-way coupled models allow the wave activity to feed back upon the atmosphere. The European Centre for Medium-Range Weather Forecasts (ECMWF) coupled atmosphere-wave forecast system described below facilitates this through exchange of the Charnock parameter which controls the sea surface roughness. This allows the atmosphere to respond to changes in the surface roughness as the wind sea builds up or decays.

EXAMPLES

WAVEWATCH

The operational wave forecasting systems at NOAA are based on the WAVEWATCH III model. This system has a global domain of approximately 100 km resolution, with nested regional domains for the northern hemisphere oceanic basins at approximately 25 km resolution. Physics includes wave field refraction, nonlinear resonant interactions, sub-grid representations of unresolved islands, and dynamically updated ice coverage. Wind data is provided from the GDAS data assimilation system for the GFS weather model. Up to 2008, the model was limited to regions outside the surf zone where the waves are not strongly impacted by shallow depths. The model can handle the effects of currents on waves from its early design by Hendrik Tolman in the 1990s, and is now extended for nearshore applications.

WAM

The wave model WAM was the first so-called third generation prognostic wave model where the two-dimensional wave spectrum was allowed to evolve freely (up to a cut-off frequency) with no constraints on the spectral shape. The model underwent a series of software updates from its inception in the late 1980s. The last official release is Cycle 4.5, maintained by the German Helmholtz Zentrum, Geesthacht.

ECMWF has incorporated WAM into its deterministic and ensemble forecasting system., known as the Integrated Forecast System (IFS). The model currently comprises 36 frequency bins and 36 propagation directions at an average spatial resolution of 25 km. The model has been coupled to the atmospheric component of IFS since 1998.

Other models

Wind wave forecasts are issued regionally by Environment Canada. Regional wave predictions are also produced by universities, such as Texas A&M University's use of the SWAN model (developed by Delft University of Technology) to forecast waves in the Gulf of Mexico. Other wind wave models include the U.S. Navy Standard Surf Model (NSSM). The private sector is also active in producing wind wave simulations and surf forecasts. For example, Oceanweather Inc. provides global operational forecasts and hindcasts of the sea state.

VALIDATION

Comparison of the wave model forecasts with observations is essential for characterizing model deficiencies and identifying areas for improvement. In-situ observations are obtained from buoys, ships and oil platforms. Altimetry data from satellites, such as GEOSAT and TOPEX, can also be used to infer the characteristics of wind waves. Hindcasts of wave models during extreme conditions also serves as a useful test bed for the models.

REANALYSES

A retrospective analysis, or reanalysis, combines all available observations with a physical model to describe the state of a system over a time period of decades. Wind waves are a part of both the NCEP Reanalysis and the ERA-40 from the ECMWF. Such resources permit the creation of monthly wave climatologies, and can track the variation of wave activity on interannual and multi-decadal time scales. During the northern hemisphere winter, the most intense wave activity is located in the central North Pacific south of the Aleutians, and in the central North Atlantic south of Iceland. During the southern hemisphere winter, intense wave activity circumscribes the pole at around 50°S, with 5 m significant wave heights typical in the southern Indian Ocean.

7

Semi-submersible

A semi-submersible (semisubmerged ship) is a specialised marine vessel used in a number of specific offshore roles such as offshore drilling rigs, safety vessels, oil production platforms, and heavy lift cranes. They are designed with good stability and seakeeping characteristics. Other terms include semisubmersible, semi-sub, or simply semi.

CHARACTERISTICS

Offshore drilling in water depth greater than around 520 meters requires that operations be carried out from a floating vessel, as fixed structures are not practical. Initially in the early 1950s monohull ships were used like CUSS I, but these were found to have significant heave, pitch and yaw motions in large waves, and the industry needed more stable drilling platforms. A semi-submersible obtains its buoyancy from ballasted, watertight pontoons located below the ocean surface and wave action.

The operating deck can be located high above the sea level due to the good stability of the design, and therefore the operating deck is kept well away from the waves. Structural columns connect the pontoons and operating deck. With its hull structure submerged at a deep draft, the semi-submersible is less affected by wave loadings than a normal ship. With a small water-plane area, however, the semi-submersible is sensitive to load changes, and therefore must be carefully trimmed to maintain stability. Unlike a submarine or submersible, during normal operations, a semi-submersible vessel is never entirely underwater.

A semi-submersible vessel is able to transform from a deep to a shallow draft by deballasting (removing ballast water from the hull), and thereby become a surface vessel. The heavy lift vessels use this capability to submerge the majority of their structure, locate beneath another floating vessel, and then deballast to pick up the other vessel as a cargo.

EARLY HISTORY

The semi-submersible design was first developed for offshore drilling activities. Bruce Collipp of Shell is regarded as the inventor. But Edward

Robert Armstrong may have paved the way with his idea of "seadrome" landing strips for airplanes in the late 1920s, since his idea involved the same use of columns on ballast tanks below the surface and anchored to the ocean floor by steel cables. When oil drilling moved into offshore waters, fixed platform rigs and submersible rigs were built, but were limited to shallow waters. When demands for drilling equipment was needed in water depths greater than 100 feet (30 m) in the Gulf of Mexico, the first jackup rigs were built.

The first semisubmersible arrived by accident in 1961. Blue Water Drilling Company owned and operated the four column submersible drilling rig Blue Water Rig No.1 in the Gulf of Mexico for Shell Oil Company. As the pontoons were not sufficiently buoyant to support the weight of the rig and its consumables, it was towed between locations at a draught mid way between the top of the pontoons and the underside of the deck.

It was observed that the motions at this draught were very small, and Blue Water Drilling and Shell jointly decided that the rig could be operated in the floating mode. The first purpose built drilling semi-submersible Ocean Driller was launched in 1963. Since then, many semi-submersibles have been purpose-designed for the drilling industry mobile offshore fleet. The industry quickly accepted the semi-submersible concept and the fleet increased rapidly to 30 units by 1972.

CLASSIFICATION

Drilling rig construction has historically occurred in boom periods and therefore "batches" of drilling rigs have been built. Offshore drilling rigs have been classified in nominal "generations" depending upon the year built and water depth capability as follows:

Generation	Water Depth		Dates
First	about 600 ft	200 m	Early 1960s
Second	about 1000 ft	300 m	1969-1974
Third	about 1500 ft	500 m	Early 1980s
Fourth	about 3000 ft	1000 m	1990s
Fifth	about 7500 ft	2500 m	1998-2004
Sixth	about 10000 ft	3000 m	2005-2010

APPLICATIONS

MOBILE OFFSHORE DRILLING UNITS (MODU)

Semi-submersible rigs make stable platforms for drilling for offshore oil and gas. They can be towed into position by a tugboat and anchored, or moved by and kept in position by their own azimuth thrusters with dynamic positioning. The IMO MODU Code is an accredited design and operational guideline for Mobile Offshore Drilling Units of the semi-submersible type.

SEMI-SUBMERSIBLE CRANE VESSELS (SSCV)

The advantages of the semi-submersible vessel stability were soon recognized for offshore construction when in 1978 Heerema Marine Contractors constructed the two sister crane vessels called Balder and Hermod. These semi-submersible crane vessels (SSCV) consist of two lower hulls (pontoons), three columns on each pontoon and an upper hull. Shortly after J. Ray McDermott and Saipem also introduced SSCV's, resulting in two new enormous vessels DB-102 (now Thialf) and Saipem 7000, capable of lifting respectively 14,200 and 14,000 tons. During transit an SSCV will be de-ballasted to a draught where only part of the lower hull is submerged. During lifting operations, the vessel will be ballasted down. This way, the lower hull is well submerged. This reduces the effect of waves and swell. High stability is obtained by placing the columns far apart. The high stability allows them to lift extreme high loads.

OFFSHORE SUPPORT VESSELS (OSV)

Semi-submersibles are particularly suited to a number of offshore support vessel roles because of their good stability, large deck areas, and variable deck load (VDL). Some of the most prominent vessels are;

- Uncle John - Diving / Construction support vessel, built for Houlder Offshore in 1977
- Seaway Swan - Diving Support Vessel built in 1977
- Tharos - Offshore Safety support vessel, built in 1979 and since converted into a drilling vessel, and rechristened Transocean Marianas.
- Stadive - Diving Support Vessel (DSV) built for Shell in 1982
- Iolair - Offshore safety support vessel, built for BP in 1982, Sedco/ Phillips SS was the first built to Red Adair's recommendations. Iolair followed.
- Safe Karinia - Offshore operations vessel, built in 1982
- Polyconfidence - Offshore accommodation platform, built in 1988.
- Q4000 - Offshore Multiservice Vessel, built for Caldive in 2002
- Ocean Odyssey - Converted semi-submersible drilling rig used as a rocket launch pad.

OFFSHORE PRODUCTION PLATFORMS

When oil fields were first developed in offshore locations, drilling semi-submersibles were converted for use as combined drilling and production platforms. These vessels offered very stable and cost effective platforms. The first semi-submersible floating production platform was the Argyll FPF converted from the Transworld 58 drilling semi-submersible in 1975 for the Hamilton Brothers North Sea Argyll oil field. As the oil industry has progressed into deeper water and harsh environments, purpose-built production semi-submersible platforms were designed. The first purpose-built semi-submersible production platform was for the Balmoral field, UK North Sea in 1986.

8

Oceanography

Oceanography (compound of the Greek words ??????? meaning "ocean" and ????? meaning "to write"), also known as oceanology and marine science, is the branch of Earth science that studies the ocean. It covers a wide range of topics, including marine organisms and ecosystem dynamics; ocean currents, waves, and geophysical fluid dynamics; plate tectonics and the geology of the sea floor; and fluxes of various chemical substances and physical properties within the ocean and across its boundaries. These diverse topics reflect multiple disciplines that oceanographers blend to further knowledge of the world ocean and understanding of processes within: astronomy, biology, chemistry, climatology, geography, geology, hydrology, meteorology and physics.

HISTORY

BEFORE THE 19TH CENTURY

Humans first acquired knowledge of the waves and currents of the seas and oceans in pre-historic times. Observations on tides were recorded by Aristotle and Strabo. Early modern exploration of the oceans was primarily for cartography and mainly limited to its surfaces and of the animals that fishermen brought up in nets, though depth soundings by lead line were taken.

Although Juan Ponce de León in 1513 first identified the Gulf Stream, and the current was well-known to mariners, Benjamin Franklin made the first scientific study of it and gave it its name. Franklin measured water temperatures during several Atlantic crossings and correctly explained the Gulf Stream's cause. Franklin and Timothy Folger printed the first map of the Gulf Stream in 1769-1770.

When Louis Antoine de Bougainville, who voyaged between 1766 and 1769, and James Cook, who voyaged from 1768 to 1779, carried out their explorations in the South Pacific, information on the oceans themselves formed part of the reports. James Rennell wrote the first scientific textbooks about currents in the Atlantic and Indian oceans during the late 18th and at the beginning of 19th century.

IN THE 19TH CENTURY

Sir James Clark Ross took the first modern sounding in deep sea in 1840, and Charles Darwin published a paper on reefs and the formation of atolls as a result of the second voyage of HMS Beagle in 1831-6. Robert FitzRoy published a report in four volumes of the three voyages of the Beagle. In 1841-1842 Edward Forbes undertook dredging in the Aegean Sea that founded marine ecology.

As first superintendent of the United States Naval Observatory (1842-1861) Matthew Fontaine Maury devoted his time to the study of marine meteorology, navigation, and charting prevailing winds and currents. His Physical Geography of the Sea, 1855 was the first textbook of oceanography. Many nations sent oceanographic observations to Maury at the Naval Observatory, where he and his colleagues evaluated the information and gave the results worldwide distribution. After the middle of the 19th century, scientific societies were processing a flood of new terrestrial botanical and zoological information.

In 1871, under the recommendations of the Royal Society of London, the British government sponsored an expedition to explore world's oceans and conduct scientific investigations. Under that sponsorship, the Scots Charles Wyville Thompson and Sir John Murray launched the Challenger expedition (1872-1876). The results of this were published in 50 volumes covering biological, physical and geological aspects. 4417 new species were discovered. Other European and American nations also sent out scientific expeditions (as did private individuals and institutions). The first purpose built oceanographic ship, the "Albatros" was built in 1882. In 1893, Fridtjof Nansen allowed his ship "Fram" to be frozen in the Arctic ice. As a result he was able to obtain oceanographic data as well as meteorological and astronomical data.

IN THE 20TH CENTURY

Between 1907 and 1911 Otto Krümmel published the Handbuch der Ozeanographie, influential in awakening public interest in oceanography. The four-month 1910 North Atlantic expedition headed by Sir John Murray and Johan Hjort was at that time the most ambitious research oceanographic and marine zoological project ever, and led to the classic 1912 book The Depths of the Ocean.

The first acoustic measurement of sea depth was made in 1914. Between 1925 and 1927 the "Meteor" expedition gathered 70,000 ocean depth measurements using an echo sounder, surveying the Mid atlantic ridge. The Great Global Rift, running along the Mid Atlantic Ridge, was discovered by Maurice Ewing and Bruce Heezen in 1953 while the mountain range under the Arctic was found in 1954 by the Arctic Institute of the USSR. The theory of seafloor spreading was developed in 1960 by Harry Hammond Hess. The Ocean Drilling Project started in 1966. Deep sea vents were discovered in 1977 by John Corlis and Robert Ballard in the submersible "Alvin". In 1939 the Carnegie Institution assigned Dr. Charles S. Piggot the mission of exploring the sea bed for Radium deposits. For this deep water mission the Western Union Cable vessel the Lord Kelvin was converted to the ability to lower a cable down several miles, which had a unit attached that fired a hollow dart into the sea bed floor which was then raised with the seabed sample for analysis.

In the 1950s, Auguste Piccard invented the bathyscaphe and used the "Trieste" to investigate the ocean's depths. The nuclear submarine Nautilus made the first journey under the ice to the North Pole in 1958. In 1962 there was the first deployment of FLIP (Floating Instrument Platform), a 355 foot spar buoy. Then, in 1966, the U.S. Congress created a National Council for Marine Resources and Engineering Development. NOAA was put in charge of exploring and studying all aspects of Oceanography in the USA. It also enabled the National Science Foundation to award Sea Grant College funding to multi-disciplinary researchers in the field of oceanography.

From the 1970s, there has been much emphasis on the application of large scale computers to oceanography to allow numerical predictions of ocean conditions and as a part of overall environmental change prediction. An oceanographic buoy array was established in the Pacific to allow prediction of El Niño events.

In 1942, Sverdrup and Fleming published "The Ocean" which was a major landmark. "The Sea" (in three volumes covering physical oceanography, seawater and geology) edited by M.N. Hill was published in 1962 while the "Encyclopedia of Oceanography" by Rhodes Fairbridge was published in 1966. 1990 saw the start of the World Ocean Circulation Experiment (WOCE) which continued until 2002. Geosat seafloor mapping data became available in 1995.

IN THE 21ST CENTURY

In recent years studies advanced particular knowledge on ocean acidification, ocean heat content, ocean currents, ENSO, mapping of methane hydrate deposits, the carbon cycle, coastal erosion, weathering and climate feedbacks in regards to climate change interactions.

OCEANOGRAPHY

The study of the oceans is linked to understanding global climate changes, potential global warming and related biosphere concerns. The atmosphere and ocean are linked because of evaporation and precipitation as well as thermal flux (and solar insolation). Wind stress is a major driver of ocean currents while the ocean is a sink for atmospheric carbon dioxide. All these factors specially relate to the ocean's biogeochemical setup. Our planet is invested with two great oceans; one visible, the other invisible; one underfoot, the other overhead; one entirely envelopes it, the other covers about two thirds of its surface. -Matthew F. Maury, The Physical Geography of the Seas and Its Meteorology (1855)

OCEAN ACIDIFICATION

Ocean acidification describes the decrease in ocean pH, caused by anthropogenic carbon dioxide (CO2) emissions into the atmosphere.

OCEAN CURRENTS

Since the early ocean expeditions in oceanography, a major interest was the study of the ocean currents and temperature measurements. The tides, the Coriolis effect, changes in direction and strength of Wind, salinity and temperature are the main factors determining ocean currents. The thermohaline circulation (THC) thermo- referring to temperature and -haline referring to salt content connects 4 of 5 ocean basins and is primarily dependent on the density of sea water. Ocean currents such as the Gulf Stream are wind-driven surface currents.

OCEAN HEAT CONTENT

Oceanic heat content (OHC) is the term used to refer to the heat stored in the ocean. The changes in the ocean heat play an important role in sea level rise, because of thermal expansion. It is with high confidence that Ocean warming accounts for 90% of the energy accumulation from global warming between 1971 and 2010.

BRANCHES

The study of oceanography is divided into branches:

- Biological oceanography investigates the ecology of marine organisms in the context of the physical, chemical, and geological characteristics of their ocean environment. It is closely aligned with marine biology, though the latter has more emphasis on the biology of individual marine organisms.
- Chemical oceanography, or marine chemistry, is the study of the chemistry of the ocean and its chemical interaction with the atmosphere;
- Geological oceanography, or marine geology, is the study of the geology of the ocean floor including plate tectonics and paleoceanography;
- Physical oceanography, or marine physics, studies the ocean's physical attributes including temperature-salinity structure, mixing, waves, internal waves, surface tides, internal tides, and currents.

These branches reflect the fact that many oceanographers are first trained in the exact sciences or mathematics and then focus on applying their interdisciplinary knowledge, skills and abilities to oceanography.

Data derived from the work of Oceanographers is used in marine engineering, in the design and building of oil platforms, ships, harbours, and other structures that allow us to use the ocean safely. Oceanographic data management is the discipline ensuring that oceanographic data both past and present are available to researchers.

OCEANOGRAPHIC INSTITUTIONS

The first international organization of oceanography was created in 1902 as the International Council for the Exploration of the Sea.

In 1892 in the Scripps Institution of Oceanography was founded, Woods Hole Oceanographic Institution in 1930, Virginia Institute of Marine Science in 1938, Lamont-Doherty Earth Observatoryat Columbia University, and the School of Oceanography at University of Washington. In Britain, there is a major research institution:National Oceanography Centre, Southampton which is the successor to the Institute of Oceanography. In Australia, CSIRO Marine and Atmospheric Research, known as CMAR, is a leading center. In 1921 the International Hydrographic Bureau (IHB) was formed in Monaco.

CHEMICAL OCEANOGRAPHY

Chemical oceanography is the study of ocean chemistry: the behavior of the chemical elements within the Earth's oceans. The ocean is unique in that it contains - in greater or lesser quantities - nearly every element in the periodic table. Much of chemical oceanography describes the cycling of these elements both within the ocean and with the other spheres of the Earth system (see biogeochemical cycle).

These cycles are usually characterised as quantitative fluxes between constituent reservoirs defined within the ocean system and as residence times within the ocean. Of particular global and climatic significance are the cycles of the biologically active elements such as carbon, nitrogen, and phosphorus as well as those of some important trace elements such as iron.

Another important area of study in chemical oceanography is the behaviour of isotopes (see isotope geochemistry) and how they can be used as tracers of past and present oceanographic and climatic processes. For example, the incidence of O (the heavy isotope of oxygen) can be used as an indicator of polar ice sheet extent, and boron isotopes are key indicators of the pH and CO2 content of oceans in the geologic past.

PHYSICAL OCEANOGRAPHY

Physical oceanography is the study of physical conditions and physical processes within the ocean, especially the motions and physical properties of ocean waters. Physical oceanography is one of several sub-domains into which oceanography is divided. Others include biological, chemical and geological oceanographies.

PHYSICAL SETTING

The pioneering oceanographer Matthew Maury said in 1855 "Our planet is invested with two great oceans; one visible, the other invisible; one underfoot, the other overhead; one entirely envelopes it, the other covers about two thirds of its surface." The fundamental role of the oceans in shaping Earth is acknowledged by ecologists, geologists, meteorologists, climatologists, geographers and others interested in the physical world. An Earth without oceans would truly be unrecognizable.

Roughly 97% of the planet's water is in its oceans, and the oceans are the source of the vast majority of water vapor that condenses in the atmosphere and falls as rain or snow on the continents. The tremendous heat capacity of the oceans moderates the planet's climate, and its absorption of various gases affects the composition of the atmosphere. The ocean's influence extends even to the composition of volcanic rocks through seafloor metamorphism, as well as to that of volcanic gases and magmas created at subduction zones. The oceans are far deeper than the continents are tall; examination of the Earth's hypsographic curve shows that the average elevation of Earth's landmasses is only 840 metres (2,760 ft), while the ocean's average depth is 3,800 metres (12,500 ft). Though this apparent discrepancy is great, for both land and sea, the respective extremes such as mountains and trenches are rare.

TEMPERATURE, SALINITY AND DENSITY

Because the vast majority of the world ocean's volume is deep water, the mean temperature of seawater is low; roughly 75% of the ocean's volume has a temperature from 0° - 5°C (Pinet 1996). The same percentage falls in a salinity range between 34-35 ppt (3.4-3.5%) (Pinet 1996). There is still quite a bit of variation, however. Surface temperatures can range from below freezing near the poles to 35°C in restricted tropical seas, while salinity can vary from 10 to 41 ppt (1.0-4.1%).

The vertical structure of the temperature can be divided into three basic layers, a surface mixed layer, where gradients are low, a thermocline where gradients are high, and a poorly stratified abyss. In terms of temperature, the ocean's layers are highly latitude-dependent; the thermocline is pronounced in the tropics, but nonexistent in polar waters (Marshak 2001). The halocline usually lies near the surface, where evaporation raises salinity in the tropics, or meltwater dilutes it in polar regions. These variations of salinity and temperature with depth change the density of the seawater, creating the pycnocline.

CIRCULATION

Energy for the ocean circulation (and for the atmospheric circulation) comes from solar radiation and gravitational energy from the sun and moon. The amount of sunlight absorbed at the surface varies strongly with latitude, being greater at the equator than at the poles, and this engenders fluid motion in both the atmosphere and ocean that acts to redistribute heat from the equator towards the poles, thereby reducing the temperature gradients that would exist in the absence of fluid motion. Perhaps three quarters of this heat is carried in the atmosphere; the rest is carried in the ocean.

The atmosphere is heated from below, which leads to convection, the largest expression of which is the Hadley circulation. By contrast the ocean is heated from above, which tends to suppress convection. Instead ocean deep water is formed in polar regions where cold salty waters sink in fairly restricted areas. This is the beginning of the thermohaline circulation. Oceanic currents are largely driven by the surface wind stress; hence the large-scale atmospheric circulation is important to understanding the ocean circulation. The Hadley circulation leads to Easterly winds in the tropics and Westerlies in mid-latitudes. This leads to slow equatorward flow throughout most of a subtropical ocean basin (the Sverdrup balance). The return flow occurs in an intense, narrow, poleward western boundary current. Like the atmosphere, the ocean is far wider than it is deep, and hence horizontal motion is in general much faster than vertical motion. In the southern hemisphere there is a continuous belt of ocean, and hence the mid-latitude westerlies force the strong Antarctic Circumpolar Current. In the northern hemisphere the land masses prevent this and the ocean circulation is broken into smaller gyres in the Atlantic and Pacific basins.

CORIOLIS EFFECT

The Coriolis effect results in a deflection of fluid flows (to the right in the Northern Hemisphere and left in the Southern Hemisphere). This has profound effects on the flow of the oceans. In particular it means the flow goes around high and low pressure systems, permitting them to persist for long periods of time. As a result, tiny variations in pressure can produce measurable currents. A slope of one part in one million in sea surface height, for example, will result in a current of 1 cm/s at mid-latitudes. The fact that the Coriolis effect is largest at the poles and weak at the equator results in sharp, relatively steady western boundary currents which are absent on eastern boundaries.

EKMAN TRANSPORT

Ekman transport results in the net transport of surface water 90 degrees to the right of the wind in the Northern Hemisphere, and 90 degrees to the left of the wind in the Southern Hemisphere. As the wind blows across the surface of the ocean, it "grabs" onto a thin layer of the surface water. In turn, that thin sheet of water transfers motion energy to the thin layer of water under it, and so on. However, because of the Coriolis Effect, the direction of travel of the layers of water slowly move farther and farther to the right as they get deeper in the Northern Hemisphere, and to the left in the Southern Hemisphere. In most cases, the very bottom layer of water affected by the wind is at a depth of 100 m - 150 m and is traveling about 180 degrees, completely opposite of the direction that the wind is blowing. Overall, the net transport of water would be 90 degrees from the original direction of the wind.

LANGMUIR CIRCULATION

Langmuir circulation results in the occurrence of thin, visible stripes, called windrows on the surface of the ocean parallel to the direction that the wind is blowing. If the wind is blowing with more than 3 m s, it can create parallel windrows alternating upwelling and downwelling about 5-300 m apart. These windrows are created by adjacent ovular water cells (extending to about 6 m (20 ft) deep) alternating rotating clockwise and counterclockwise. In the convergence zones debris, foam and seaweed accumulates, while at the divergence zones plankton are caught and carried to the surface. If there are many plankton in the divergence zone fish are often attracted to feed on them.

OCEAN-ATMOSPHERE INTERFACE

At the ocean-atmosphere interface, the ocean and atmosphere exchange fluxes of heat, moisture and momentum.

Heat

The important heat terms at the surface are the sensible heat flux, the latent heat flux, the incoming solar radiation and the balance of long-wave (infrared) radiation. In general, the tropical oceans will tend to show a net gain of heat, and the polar oceans a net loss, the result of a net transfer of energy polewards in the oceans. The oceans' large heat capacity moderates the climate of areas adjacent to the oceans, leading to a maritime climate at such locations. This can be a result of heat storage in summer and release in winter; or of transport of heat from warmer locations: a particularly notable example of this is Western Europe, which is heated at least in part by the north atlantic drift.

Momentum

Surface winds tend to be of order meters per second; ocean currents of order centimeters per second. Hence from the point of view of the atmosphere, the ocean can be considered effectively stationary; from the point of view of the ocean, the atmosphere imposes a significant wind stress on its surface, and this forces large-scale currents in the ocean. Through the wind stress, the wind generates ocean surface waves; the longer waves have a phase velocity tending towards the wind speed. Momentum of the surface winds is transferred into the energy flux by the ocean surface waves. The increased roughness of the ocean surface, by the presence of the waves, changes the wind near the surface.

Moisture

The ocean can gain moisture from rainfall, or lose it through evaporation. Evaporative loss leaves the ocean saltier; the Mediterranean and Persian Gulf for example have strong evaporative loss; the resulting plume of dense salty water may be traced through the Straits of Gibraltar into the Atlantic Ocean. At one time, it was believed that evaporation/precipitation was a major driver of ocean currents; it is now known to be only a very minor factor.

PLANETARY WAVES

A Kelvin wave is any progressive wave that is channeled between two boundaries or opposing forces (usually between the Coriolis force and a coastline or the equator). There are two types, coastal and equatorial. Kelvin waves are gravity driven and non-dispersive.

This means that Kelvin waves can retain their shape and direction over long periods of time. They are usually created by a sudden shift in the wind, such as the change of the trade winds at the beginning of the El Niño-Southern Oscillation. Coastal Kelvin waves follow shorelines and will always propagate in a counterclockwise direction in the Northern hemisphere (with the shoreline to the right of the direction of travel) and clockwise in the Southern hemisphere.

Equatorial Kelvin waves propagate to the east in the Northern and Southern hemispheres, using the equator as a guide. Kelvin waves are known to have very high speeds, typically around 2-3 meters per second. They have wavelengths of thousands of kilometers and amplitudes in the tens of meters.

Rossby Waves

Rossby waves, or planetary waves are huge, slow waves generated in the troposphere by temperature differences between the ocean and the continents. Their major restoring force is the change in Coriolis force with latitude. Their wave amplitudes are usually in the tens of meters and very large wavelengths. They are usually found at low or mid latitudes There are two types of Rossby waves, barotropic and baroclinic. Barotropic Rossby waves have the highest speeds and do not vary vertically. Baroclinic Rossby waves are much slower.

The special identifying feature of Rossby waves is that the phase velocity of each individual wave always has a westward component, but the group velocity can be in any direction. Usually the shorter Rossby waves have an eastward group velocity and the longer ones have a westward group velocity.

CLIMATE VARIABILITY

The interaction of ocean circulation, which serves as a type of heat pump, and biological effects such as the concentration of carbon dioxide can result in global climate changes on a time scale of decades. Known climate oscillations resulting from these interactions, include the Pacific decadal oscillation, North Atlantic oscillation, and Arctic oscillation. The oceanic process of thermohaline circulation is a significant component of heat redistribution across the globe, and changes in this circulation can have major impacts upon the climate.

ANTARCTIC CIRCUMPOLAR WAVE

This is a coupled ocean/atmosphere wave that circles the Southern Ocean about every eight years. Since it is a wave-2 phenomenon (there are two peaks and two troughs in a latitude circle) at each fixed point in space a signal with a period of four years is seen. The wave moves eastward in the direction of the Antarctic Circumpolar Current.

OCEAN CURRENTS

Among the most important ocean currents are the:

- Antarctic Circumpolar Current
- Deep ocean (density-driven)
- Western boundary currents
  - Gulf Stream
  - Kuroshio Current
  - Labrador Current
  - Oyashio Current
  - Agulhas Current
  - Brazil Current
  - East Australia Current
- Eastern Boundary currents
  - California Current
  - Canary Current
  - Peru Current
  - Benguela Current

Further information: Ocean gyre

ANTARCTIC CIRCUMPOLAR

The ocean body surrounding the Antarctic is currently the only continuous body of water where there is a wide latitude band of open water. It interconnects the Atlantic, Pacific and Indian oceans, and provide an uninterrupted stretch for the prevailing westerly winds to significantly increase wave amplitudes. It is generally accepted that these prevailing winds are primarily responsible for the circumpolar current transport. This current is now thought to vary with time, possibly in an oscillatory manner.

DEEP OCEAN

In the Norwegian Sea evaporative cooling is predominant, and the sinking water mass, the North Atlantic Deep Water (NADW), fills the basin and spills southwards through crevasses in the submarine sills that connect Greenland, Iceland and Britain.

It then flows along the western boundary of the Atlantic with some part of the flow moving eastward along the equator and then poleward into the ocean basins. The NADW is entrained into the Circumpolar Current, and can be traced into the Indian and Pacific basins. Flow from the Arctic Ocean Basin into the Pacific, however, is blocked by the narrow shallows of the Bering Strait.

WESTERN BOUNDARY

An idealised subtropical ocean basin forced by winds circling around a high pressure (anticyclonic) systems such as the Azores-Bermuda high develops a gyre circulation with slow steady flows towards the equator in the interior. As discussed by Henry Stommel, these flows are balanced in the region of the western boundary, where a thin fast polewards flow called a western boundary current develops. Flow in the real ocean is more complex, but the Gulf stream, Agulhas and Kuroshio are examples of such currents. They are narrow (approximately 100 km across) and fast (approximately 1.5 m/s).

Equatorwards western boundary currents occur in tropical and polar locations, e.g. the East Greenland and Labrador currents, in the Atlantic and the Oyashio. They are forced by winds circulation around low pressure (cyclonic)

Gulf stream

The Gulf Stream, together with its northern extension, North Atlantic Current, is a powerful, warm, and swift Atlantic ocean current that originates in the Gulf of Mexico, exits through the Strait of Florida, and follows the eastern coastlines of the United States and Newfoundland to the northeast before crossing the Atlantic Ocean.

Kuroshio

The Kuroshio Current is an ocean current found in the western Pacific Ocean off the east coast of Taiwan and flowing northeastward past Japan, where it merges with the easterly drift of the North Pacific Current. It is analogous to the Gulf Stream in the Atlantic Ocean, transporting warm, tropical water northward towards the polar region.

HEAT FLUX

Heat storage and transfer in the ocean is very uneven.

Sea level change

Tide gauges and satellite altimetry suggest an increase in sea level of 1.5-3 mm/yr over the past 100 years. The IPCC predicts that by 2100, global warming will lead to a sea level rise of 110 to 880 mm.

RAPID VARIATIONS

Tides

The rise and fall of the oceans due to tidal effects is a key influence upon the coastal areas. Ocean tides on the planet Earth are created by the gravitational effects of the Sun and Moon. The tides produced by these two bodies are roughly comparable in magnitude, but the orbital motion of the Moon results in tidal patterns that vary over the course of a month. The ebb and flow of the tides produce a cyclical current along the coast, and the strength of this current can be quite dramatic along narrow estuaries. Incoming tides can also produce a tidal bore along a river or narrow bay as the water flow against the current results in a wave on the surface.

Tide and Current (Wyban 1992) clearly illustrates the impact of these natural cycles on the lifestyle and livelihood of Native Hawaiians tending coastal fishponds. Aia ke ola ka hana meaning . . . Life is in labor. Tidal resonance occurs in the Bay of Fundy since the time it takes for a large wave to travel from the mouth of the bay to the opposite end, then reflect and travel back to the mouth of the bay coincides with the tidal rhythm producing the world's highest tides. As the surface tide oscillates over topography, such as submerged seamounts or ridges, it generates internal waves at the tidal frequency, which are known as internal tides.

Tsunamis

A series of surface waves can be generated due to large-scale displacement of the ocean water. These can be caused by sub-marine landslides, seafloor deformations due to earthquakes, or the impact of a large meteorite. The waves can travel with a velocity of up to several hundred km/hour across the ocean surface, but in mid-ocean they are barely detectable with wavelengths spanning hundreds of kilometers.

Tsunamis, originally called tidal waves, were renamed because they are not related to the tides. They are regarded as shallow-water waves, or waves in water with a depth less than 1/20 their wavelength. Tsunamis have very large periods, high speeds, and great wave heights. The primary impact of these waves is along the coastal shoreline, as large amounts of ocean water are cyclically propelled inland and then drawn out to sea. This can result in significant modifications to the coastline regions where the waves strike with sufficient energy. The tsunami that occurred in Lituya Bay, Alaska on July 9, 1958 was 520 m (1,710 ft) high and is the biggest tsunami ever measured, almost 90 m (300 ft) taller than the Sears Tower in Chicago and about 110 m (360 ft) taller than the former World Trade Center in New York.

SURFACE WAVES

The wind generates ocean surface waves, which have a large impact on offshore structures, ships, coastal erosion and sedimentation, as well as harbours. After their generation by the wind, ocean surface waves can travel (as swell) over long distances.

9

Deep-sea Exploration

Deep-sea exploration is the investigation of physical, chemical, and biological conditions on the sea bed, for scientific or commercial purposes. Deep-sea exploration is considered as a relatively recent human activity compared to the other areas of geophysical research, as the depths of the sea have been investigated only during comparatively recent years. The ocean depths still remain as a largely unexplored part of the planet, and form a relatively undiscovered domain.

In general, modern scientific Deep-sea exploration can be said to have begun when French scientist Pierre Simon de Laplace investigated the average depth of the Atlantic ocean by observing tidal motions registered on Brazilian and African coasts. He calculated the depth to be 3,962 m (13,000 ft), a value later proven quite accurate by soundings measurement. Later on, with increasing demand for submarine cables installment, accurate soundings was required and the first investigations of the sea bottom were undertaken. First deep-sea life forms were discovered in 1864 when Norwegian researchers obtained a sample of a stalked crinoid at a depth of 3,109 m (10,200 ft).

The British Government sent out the Challenger expedition (a ship called the HMS Challenger) in 1872 which discovered 715 new genera and 4,417 new species of marine organisms over the space of 4 years. The first instrument used for deep-sea investigation was the sounding weight, used by British explorer Sir James Clark Ross. With this instrument, he reached a depth of 3,700 m (12,140 ft) in 1840. The Challenger expedition used similar instruments called Baillie sounding machines to extract samples from the sea bed. In 1960, Jacques Piccard and US Navy Lieutenant Donald Walsh descended in the bathyscaphe Trieste into the Mariana Trench, the deepest part of the world's oceans, to make the deepest dive in history: 10,915 meters (35,810 ft). On 25 March 2012, filmmaker James Cameron descended into the deepest part of the Mariana Trench and, for the first time, is expected to have filmed and sampled the bottom.

BRIEF HISTORY

Throughout history, scientists have relied on a number of instruments to measure, map, and observe the ocean's depths. One of the first instruments used to examine the seafloor was the sounding weight. Ancient Viking sailors took measurements of sea depth and sampled seafloor sediments with this instrument, which consisted of a lead weight with a hollow bottom attached to a line. Once the weight reached the sea bottom and collected a sample of the seabed, the line was hauled back on board ship and measured in fathom. Cornelius Drebbel, a Dutch architect, is generally given credit for construction of the first submarine. His submersible boat consisted of a wooden frame sheathed in animal skin. Oars, with its openings were sealed with tight-fitting leather flaps, extended out the sides to propel the craft through the water, at depths up to 4.6 metres (15 ft). Drebel tested his submarine in the Thames River in England in sometime between 1620 and 1624. It is believed that King James I may have enjoyed a short ride in the craft.

However, the nature of the deep ocean remained an unrevealed mystery until the mid-19th century. Scientists and artists alike imagined the deep sea as a lifeless soup of placid water. French author Jules Verne, who helped pioneer the science-fiction genre, portrayed the deep ocean as contained in a bowl of static rock in his "Twenty Thousand Leagues under the Sea". By the late 1860s, controversial modern scientific theories, the origin of life by evolution and the enormity of geologic time had created a foundation of scientific curiosity and provoked a rising interest in marine exploration. The Royal Society of England thus initiated an ambitious oceanographic mission to expand a scarce collection of existing marine data that included Charles Darwin's observations during the voyage of the HMS Beagle (1831-1836), a bathymetric chart created by U.S. Navy Lt. Matthew Maury to aid installation of the first trans-continent telegraph cables in 1858, and a few examples of deep marine creatures.

From 1872 to 1876, a landmark ocean study was carried out by British scientists aboard HMS Challenger, a sailing vessel that was redesigned into a laboratory ship. The HMS Challenger expedition covered 127,653 km (68,890 nautical miles), and shipboard scientists collected hundreds of samples, hydrographic measurements, and specimens of marine life. They are also credited with providing the first real view of major seafloor features such as the deep ocean basins. They discovered more than 4,700 new species of marine life, including deep-sea organisms.

Deep-sea exploration advanced considerably in the 1900s thanks to a series of technological inventions, ranging from sonar system to detect the presence of objects underwater through the use of sound to manned deep-diving submersibles such as DSV Alvin. Operated by the Woods Hole Oceanographic Institution, Alvin is designed to carry a crew of three people to depths of 4,000 meters (13,124 ft). The submarine is equipped with lights, cameras, computers, and highly maneuverable robotic arms for collecting samples in the darkness of the ocean's depths. However, the voyage to the ocean bottom is still a challenging experience. Scientists are working to find ways to study this extreme environment from the shipboard. With more sophisticated use of fiber optics, satellites, and remote-control robots, scientists one day may explore the deep sea from a computer screen on the deck rather than out of a porthole.

MILESTONES OF DEEP SEA EXPLORATION

The extreme conditions in the deep sea require elaborate methods and technologies, which has been the main reason why its exploration has a comparatively short history. In the following, important key stones of deep sea exploration are listed.

- 1521: Ferdinand Magellan tried to measure the depth of the Pacific ocean with a 2400 ft weighted line, but did not find bottom.
- 1818: The British researcher Sir John Ross was the first to find that the deep sea is inhabited by life when catching jellyfish and worms in about 2000 m (6550 ft) depth with a special device.
- 1843: Nevertheless, Edward Forbes claimed that diversity of life in the deep sea is little and decreases with increasing depth. He stated that there could be no life in waters deeper than 550 m (1800 ft), the so-called Abyssus Theory.
- 1850: Near the Lofoten, Michael Sars found a rich deep sea fauna in a depth of 800 m (2600 ft) thereby refuting the Abyssus Theory.
- 1872-1876: The first systematic deep sea exploration was conducted by the Challenger Expedition on board the ship HMS Challenger led by Charles Wyville Thomson. This expedition revealed that the deep sea harbours a diverse, specialized biota.
- 1890-1898: First Austrian-Hungarian deep sea expedition on board the ship SMS Pola led by Franz Steindachner in the eastern Mediterranean and the Red Sea.
- 1898-1899: First German deep sea expedition on board the ship Valdivia led by Carl Chun; found many new species from depths greater than 4000 m (13000 ft) in the southern Atlantic Ocean.
- 1930: William Beebe and Otis Barton are the first humans to reach the Deep Sea when diving in the so-called Bathysphere, made from steel. They reach a depth of 435 m (1430 ft), where they observed jellyfish and shrimp.
- 1934: The Bathysphere reached a depth of 923 m (3028 ft).
- 1948: Otis Barton set out for a new record reaching a depth of 1370 m (4495 ft).
- 1960: Jacques Piccard and Don Walsh reached the bottom of the Challenger Deep in the Mariana Trench, descending to a depth of 10,740 m (35236 ft) in their deep sea vessel Trieste, where they observed fish and other deep sea organisms.
- 2012: The vessel Deepsea Challenger, piloted by James Cameron,

completes the second manned voyage and first solo mission to the bottom of the Challenger Deep.

OCEANOGRAPHIC INSTRUMENTATION

The sounding weight, one of the first instruments used for the sea bottom investigation, was designed as a tube on the base which forced the seabed in when it hit the bottom of the ocean. British explorer Sir James Clark Ross fully employed this instrument to reach a depth of 3,700 m (12,140 ft) in 1840. The sounding weights used on the HMS Challenger were slightly advanced called "Baillie sounding machine". The British researchers used wire-line soundings to investigate sea depths and collected hundreds of biological samples from all the oceans except the Arctic. Also used on the HMS Challenger were dredges and scoops, suspended on ropes, with which samples of the sediment and biological specimens of the seabed could be obtained.

A more advanced version of the sounding weight is the gravity corer. The gravity corer allows researchers to sample and study sediment layers at the bottom of oceans. The corer consists of an open-ended tube with a lead weight and a trigger mechanism that releases the corer from its suspension cable when the corer is lowered over the seabed and a small weight touches the ground. The corer falls into the seabed and penetrates it to a depth of up to 10 m (33 ft). By lifting the corer, a long, cylindrical sample is extracted in which the structure of the seabed's layers of sediment is preserved. Recovering sediment cores allows scientists to see the presence or absence of specific fossils in the mud that may indicate climate patterns at times in the past, such as during the ice ages. Samples of deeper layers can be obtained with a corer mounted in a drill. The drilling vessel JOIDES Resolution is equipped to extract cores from depths of as much as 1,500 m (4900 ft) below the ocean bottom. Echo-sounding instruments have also been widely used to determine the depth of the sea bottom since World War II.

This instrument is used primarily for determining the depth of water by means of an acoustic echo. A pulse of sound sent from the ship is reflected from the sea bottom back to the ship, the interval of time between transmission and reception being proportional to the depth of the water. By registering the time lapses between outgoing and returning signals continuously on paper tape, a continuous mapping of the seabed is obtained. The majority of the ocean floor has been mapped in this way.

In addition, high-resolution television cameras, thermometers, pressure meters, and seismographs are other notable instruments for deep-sea exploration invented by the technological advance. These instruments are either lowered to the sea bottom by long cables or directly attached to submersible buoys. Deep-sea currents can be studied by floats carrying an ultrasonic sound device so that their movements can be tracked from aboard the research vessel. Such vessels themselves are equipped with state -of-art navigational instruments, such as satellite navigation systems, and global positioning systems that keep the vessel in a live position relative to a sonar beacon on the bottom of the ocean.

OCEANOGRAPHIC SUBMERSIBLES

Because of the high pressure, the depth to which a diver can descend without special equipment is limited. The deepest recorded made by a skin diver is 127 meters (417 ft). The deepest record made by a scuba diver is not much deeper, at 145 meters (475 ft). Revolutionary new diving suits, such as the "JIM suit," allows divers to reach depths up to approximately 600 meters (2,000 ft). Some additional suits feature thruster packs that boost a diver to different locations underwater.

To explore even deeper depths, deep-sea explorers must rely on specially constructed steel chambers to protect them. The American explorer William Beebe, also a naturalist from Columbia University in New York, was the designer of the first practical bathysphere to observe marine species at depths that could not be reached by a diver. The Bathysphere, a spherical steel vessel, was designed by Beebe and his fellow engineer Otis Barton, an engineer at Harvard University. In 1930 Beebe and Barton reached a depth of 435 m (about 1425 ft), and 923 m (3028 ft) in 1934. The potential danger was that if the cable broke, the occupants could not return to the surface. During the dive, Beebe peered out of a porthole and reported his observations by telephone to Barton who was on the surface.

In 1948, Swiss physicist Auguste Piccard tested a much deeper-diving vessel he invented called the bathyscaphe, a navigable deep-sea vessel with its gasoline-filled float and suspended chamber or gondola of spherical steel. On an experimental dive in the Cape Verde Islands, his bathyscaphe successfully withstood the pressure on it at 1,402 meters (4,600 ft), but its body was severely damaged by heavy waves after the dive. In 1954, with this bathyscaphe, Piccard reached a depth of 4,000 m (13,125 ft). In 1953, his son Jacques Piccard joined in building new and improved bathyscaphe Trieste, which dived to 3,139 meters (10,300 ft) in field trials. The U.S. Navy acquired Trieste in 1958 and equipped it with a new cabin to enable it to reach deep ocean trenches. In 1960, Jacques Piccard and Navy Lieutenant Donald Walsh descended in Trieste to the deepest known point on Earth - the Challenger Deep in the Mariana Trench, successfully making the deepest dive in history: 10,915 meters (35,810 ft).

An increasing number of occupied submersibles are now employed around the world. The American-built DSV Alvin that is operated by the Woods Hole Oceanographic Institution, is a three-person submarine that can dive to about 3,600 m (12,000 ft) and is equipped with a mechanical manipulator to collect bottom samples. Alvin made its first test dive in 1964, and has performed more than 3,000 dives to average depths of 1,829 meters

(6,000 ft). Alvin has also involved in a wide variety of research projects, such as one where giant tube worms were discovered on the Pacific Ocean floor near the Galápagos Islands.

UNMANNED SUBMERSIBLES

One of the first unmanned deep sea vehicles was developed by the University of California with a grant from the Alan Hancock Foundation in the early 1950s to develop a more economical method of taking photos miles under the sea with an unmanned steel high pressure 3,000 lb sphere called a benthograph which contained a camera and strobe light. The original benthograph built by USC was very successful in taking a series of underwater photos till it became wedged between some rocks and could not be retrieved.

ROVs, or Remote Operated Vehicles, are seeing increasing use in underwater exploration. These submersibles are piloted through a cable which connects to the surface ship, and they can reach depths of up to 6,000 meters. New developments in robotics have also led to the creation of AUVs, or Autonomous Underwater Vehicles. The robotic submarines are programmed in advance, and receive no instruction from the surface. HROV combine features of both ROVs and AUV, operating independently or with a cable. Argo was employed in 1985 to locate the wreck of the RMS Titanic; the smaller Jason was also used to explore the ship wreck.

SCIENTIFIC RESULTS

In 1974 the Alvin (operated by the Woods Hole Oceanographic Institution and the (Deep Sea Place Research Center), the French bathyscaphe Archimède, and the French diving saucer Cyane, assisted by support ships and the Glomar Challenger, explored the great rift valley of the Mid-Atlantic Ridge, southwest of the Azores. About 5,200 photographs of the region were taken, and samples of relatively young solidified magma were found on each side of the central fissure of the rift valley, giving additional proof that the seafloor spreads at this site at a rate of about 2.5 cm (about 1 in) per year. In a series of dives conducted between 1979-1980 into the Galápagos rift, off the coast of Ecuador, French,Italian, Mexican, and U.S. scientists found vents, nearly 9 m (nearly 30 ft) high and about 3.7 m (about 12 ft) across, discharging a mixture of hot water (up to 300°C/570°F) and dissolved metals in dark, smoke-like plumes. These hot springs play an important role in the formation of deposits that are enriched in copper, nickel, cadmium, chromium, and uranium.

OCEANIC PHYSICAL-BIOLOGICAL PROCESS

Due to the higher density of sea water (1,030 kg m) than air (1.2 kg m), the force exerted by the same velocity on an organism is 827 times stronger in the ocean. When waves crash on the shore, the force exerted on littoral organisms can be equivalent to several tons.

ROLES OF WATER

Water forms the ocean, produces the high density fluid environment and greatly affects the oceanic organisms.

1. Sea water produces buoyancy and provides support for plants and animals. That's the reason why in the ocean organisms can be that huge like the blue whale and macrophytes. And the densities or rigidities of the oceanic organisms are relative low compared with that of the terrestrial species. The water environment allows the organism to be soft, watery and huge. To be watery and transparent is a successful way to avoid predation.

1. Sea water can prevent desiccation although it is much saltier than fresh water. For oceanic organism, not like terrestrial plants and animals, water is never a problem.
2. Sea water carries oxygen and nutrients to oceanic organisms, which allow them to be planktonic or settled. The dissolved minerals and oxygen flow with currents/circulations. Oceanic plants and animals easily capture what they need for their daily life, which make them 'lazy' and 'slow'.
3. Sea water removes waste from animals and plants. Sea water is cleaner than we can imagine. Because of the huge volume of ocean, the waste produced by oceanic organisms and even human activities can hardly get the sea water polluted. The waste is not only 'waste' but also an important food source. Bacteria remineralize and recycle the organic matter back to the main oceanic food web.
4. Sea water transport organisms, which facilitates the food capture and fertilization. Many settled bottom organisms use their tentacles to catch planktonic food.

REYNOLDS NUMBER

Water flow can be described as laminar or turbulent. Laminar flow is characterized by smooth motion: neighboring particles advected by such a flow will follow similar paths. Turbulent flow is dominated by re-circulation, whorls, eddies and apparent randomness. In such a flow particles which are neighbors at one moment can find themselves widely separated later.

Reynolds number is the ratio of inertial forces to viscous forces. As the size of an organism and the strength of the current increases, inertial forces will eventually dominate, and the flow becomes turbulent (large Re). As the size and strength decrease, viscous forces eventually dominate and the flow becomes laminar (small Re).

Biologically there is an important distinction between plankton and neckton. Plankton are the aggregate of relatively passive organisms which float or drift with the currents, such as tiny algae and bacteria, small eggs and larvae of marine organisms, and protozoa and other minute predators. Nekton are the aggregate of actively swimming organisms which are able to move independently of water currents, such as shrimps, forage fish and sharks. As a rule of thumb, plankton are small and, if they swim at all, do so at biologically low Reynolds numbers (0.001 to 10), where the viscous behaviour of water dominates and reversible flows are the rule. Nekton, on the other hand, are larger and swim at biologically high Reynolds numbers (10 to 10), where inertial flows are the rule and eddies (vortices) are easily shed. Many organisms, such as jellyfish and most fish, start life as larva and other tiny members of the plankton community, swimming at low Reynolds numbers, but become nekton as they grow large enough to swim at high Reynolds numbers.

BERNOULLI'S PRINCIPLE

Bernoulli's Principle states that for an inviscid (frictionless) flow, an increase in the speed of the fluid occurs simultaneously with a decrease in pressure or a decrease in the fluid's potential energy. One result of Bernoulli's Principle is that slower moving current has higher pressure. This principle is used, for example, by some benthic suspension feeders.

These smart guys dig holes like U tubes with one end higher than the other end. Because of bottom drag, as water flows over the bottom the lower tube opening has a lower fluid speed and thus a higher pressure than the upper tube opening. The benthic suspension feeder can hide in the tube as the pressure difference between the tube ends drives water and suspended particles through the tube. The body shapes of many benthic creatures also exploit Bernoulli's Principle not only decrease the friction and drag but also to create lift when they move through the current.

DRAG

Drag is the tendency of an object to move in the direction of the flow. The magnitude of drag depends on the current velocity, the shape and size of the organism and the density of the fluid. Drag is a dissipative process which generally results in the generation of heat.

In sea water, drag can be decomposed into two different forms: skin friction and pressure drag.

1. Skin friction: just like other frictional forces, skin friction is a consequence of the relative movement between the surface of the organisms and its fluid environment. Under conditions of low Re, where viscous forces dominate, the skin friction is apparent and is more important, although it is also present under high Re conditions.
2. Pressure drag: pressure drag is a result of the pressure difference in front of, and behind, an organism. Incidentally, the shape that has the lowest pressure drag coefficient is a hollow hemisphere oriented in the direction of fluid flow. In the oceanic environment plants and settled animals have bodies that are soft and flexible in order to minimize the effects of pressure drag.

Besides being soft and flexible, organisms have other methods to minimize drag.

- Smooth skin: dolphins have little tear drops in their skin which traps some water so water flows over the water that is trapped. The skin feels soft and flaky and sheds every two hours. This helps dolphins swim through the sea water at high speed.
- Shark skin: the surface of shark skin is covered with tiny 'teeth' or dermal denticles. The shape and positioning of these denticles varies across the shark's body, altering the flow of water in a way to minimize form drag.
- Barracuda skin: Barracuda have hundreds of skin conduits which force the fluid flows to follow the parallel tubes and become laminar. Again, this arrangement decreases water drag.

OCEAN SURFACE TOPOGRAPHY

The ocean surface has highs and lows, similar to the hills and valleys of Earth's land surface depicted on a topographic map. These variations, called "ocean surface topography" or "dynamic sea surface topography" are mapped using measurements of sea surface height relative to Earth's geoid. Earth's geoid is a calculated surface of equal gravitational potential energy and represents the shape the sea surface would be if the ocean were not in motion. The height variations of ocean surface topography can be as much as two meters and are influenced by ocean circulation, ocean temperature, and salinity.

Ocean surface topography is used to map ocean currents, which move around the ocean's "hills" and "valleys" in predictable ways. A clockwise sense of rotation is found around "hills" in the northern hemisphere and "valleys" in the southern hemisphere. This is because of the Coriolis effect. Conversely, a counterclockwise sense of rotation is found around "valleys" in the northern hemisphere and "hills" in the southern hemisphere. Ocean surface topography is also used to understand how the ocean moves heat around the globe, a critical component of Earth's climate, and for monitoring changes in global sea level.

Ocean surface topography can be derived from ship-going measurements of temperature and salinity at depth. However, since 1992, a series of satellite altimetry missions, beginning with TOPEX/Poseidon and continued with Jason-1 and the Ocean Surface Topography Mission on the Jason-2 satellite have measured sea surface height directly. By combining these measurements with gravity measurements from NASA's Grace mission, scientists can determine sea surface topography to within a few centimeters.

A new satellite mission called the Surface Water Ocean Topography Mission has been proposed to make the first global survey of the topography of all of Earth's surface water--the ocean, lakes and rivers. This study is aimed to provide a comprehensive view of Earth's freshwater bodies from space and more much detailed measurements of the ocean surface than ever before.

OCEAN REANALYSIS

Ocean reanalysis is a method of combining historical ocean observations with a general ocean model (typically a computational model) driven by historical estimates of surface winds, heat, and freshwater, by way of a data assimilation algorithm to reconstruct historical changes in the state of the ocean. Historical observations are sparse and insufficient for understanding the history of the ocean and its circulation. By utilizing data assimilation techniques in combination with advanced computational models of the global ocean, researchers are able to interpolate the historical observations to all points in the ocean. This process has an analog in the construction of atmospheric reanalysis and is closely related to ocean state estimation.

CURRENT PROJECTS

A number of efforts have been initiated in recent years to apply data assimilation to estimate the physical state of the ocean, including temperature, salinity, currents, and sea level, in recent years. There are three alternative state estimation approaches. The first approach is used by the 'no-model' analyses, for which temperature or salinity observations update a first guess provided by climatological monthly estimates.

The second approach is that of the sequential data assimilation analyses, which move forward in time from a previous analysis using a numerical simulation of the evolving temperature and other variables produced by an ocean general circulation model. The simulation provides the first guess of the state of the ocean at the next analysis time, while corrections are made to this first guess based on observations of variables such as temperature, salinity, or sea level. The third approach is 4D-Var, which in the implementation described uses the initial conditions and surface forcing as control variables to be modified in order to be consistent with the observations as well as a numerical representation of the equations of motion through iterative solution of a giant optimization problem.

METHODOLOGIES

No-model approach

ISHII and LEVITUS begin with a first guess of the climatological monthly upper-ocean temperature based on climatologies produced by the NOAA National Oceanographic Data Center. The innovations are mapped onto the analysis levels. ISHII uses and alternative 3DVAR approach to do an objective mapping with a smaller decorrelation scale in midlatitudes (300 km) that elongates in the zonal direction by a factor of 3 at equatorial latitudes.

LEVITUS begins similarly to ISHII, but uses the technique of Cressman and Barnes with a homogeneous scale of 555 km to objectively map the temperature innovation onto a uniform grid.

Sequential approaches

The sequential approaches can be further divided into those using Optimal Interpolation and its more sophisticated cousin the Kalman Filter, and those using 3D-Var. Among those mentioned above, INGV and SODA use versions of Optimal Interpolation. CERFACS, GODAS, and GFDL all use 3DVar.

"To date we are unaware of any attempt to use Kalman Filter for multi-decadal ocean reanalyses." The 4-Dimensional Local Ensemble Transform Kalman Filter (4D-LETKF) has been applied to the Geophysical Fluid Dynamics Laboratory's (GFDL) Modular Ocean Model (MOM2) for a 7-year ocean reanalysis from January 1997-2004.

Variational (4D-Var) approach

One innovative attempt by GECCO has been made to apply 4D-Var to the decadal ocean estimation problem. This approach faces daunting computational challenges, but provides some interesting benefits including satisfying some conservation laws and the construction of the ocean model adjoint.

OCEAN HEAT CONTENT

Oceanic heat content (OHC) is the heat stored in the ocean. Oceanography and climatology are the science branches which study ocean heat content. The changes in the ocean heat play an important role in sea level rise, because of thermal expansion. It is with high confidence that Ocean warming accounts for 90% of the energy accumulation from global warming between 1971 and 2010.

DEFINITION AND MEASUREMENT

It is defined as:

$$H @ uc_p \int_{h2}^{h1} T(z)dz$$

u - water density, c_p- sea water specific heat capacity, h2 - bottom depth, h1 - top depth, $T(z)$ - temperature profile. OHC is computed from temperature measurements, often taken with a Nansen bottle. The ARGO float project deployed 3000 floaters around the worlds Ocean, which periodically dive to take temperature and salinity measurements. The World Ocean Database Project is the largest database for temperature profiles from all of the world's oceans.

RECENT CHANGES

Several studies in recent years, found a multidecadal increase in OHC of the deep and upper ocean regions and attribute the heat uptake to anthropogenic warming. Studies based on ARGO indicate that ocean surface winds change ocean heat vertical distribution. Especially the subtropical trade winds in the Pacific ocean have provided a mechanism for vertical heat distribution.

The effect are changes in the ocean currents, increasing the subtropical overturning, which are also related to the El Niño and La Niña phenomenon. Depending on stochastic natural variability fluctuations, during La Niña years around 30% more heat from the upper ocean layer is transported into the deeper ocean. Model studies indicate that ocean currents transport more heat into deeper layers during La Niña years, following changes in wind circulation. Years with increased ocean heat uptake have been associated with negative phases of the interdecadal Pacific oscillation (IPO). This has become of particular interests to climate scientist who use the data to estimate the ocean heat uptake.

ICE NAVIGATION

Ice navigation is more of an art than a science, and like most arts, does not fit neatly into any scheme for enhancement by electronic means. It is an art learned in theory at navigation school, and perfected by many years of practical experience in forcing ships though ice. The successful ice-navigator is living proof that the human brain processes data faster, and with more accurate results for correct decision-making, than any computer currently in service.

Thus, it becomes clear that electronic assistance to the process will consist mainly of the provision of tools, which will not only make the planning process for a voyage through ice more accurate, but can also quickly incorporate changes to the ice conditions while en route, and the consequent modifications to the base plan for that route.

The successful ice-navigator can be described as one who succeeds in bringing a vessel from departure point to destination in a reasonable amount of time, without the vessel suffering significant damage, or serious delay caused by any ice which has been met along the route. The basic premise that ice navigation will involve some risk of damage, some delay in arrival, and much extra effort on behalf of the crew, will never be negated by the provision of any electronic assistance.

Ship-owners and charterers must accept that adding electronic gadgetry is not going to be a panacea for all the problems posed when their vessels are committed to routes through ice-covered waters, and no artificial intelligence will ever be able to substitute for human experience, navigating in an infinite variety of ice conditions.

ICE AVOIDANCE - THE FORMULATION OF A STRATEGIC PLAN

For those who have never - or seldom - navigated in ice, perhaps it would be valuable to present a brief description of the process. Once the decision has been made to send a vessel between ports where sea-ice may be encountered, a Strategic Plan is formulated. A "Base Plan" of the route is selected, founded upon safe open water navigation principles. A map of the current ice conditions is overlaid on the "Base Plan" chart, and modifications to the selected courses are made, so as to avoid the ice completely if possible, or to minimise the time in serious ice conditions if not. This is the "Strategic Route". The basic principle in ice navigation is, of course, to avoid ice if you can. Even if staying in the open water extends the distance between ports by a substantial amount, the time difference can usually be made up by increased speed possible in open water.

TACTICAL PLAN NEEDED IN THE LIGHT OF DAILY CHANGING ICE CONDITIONS

While the vessel is steaming towards the destination, the ice conditions are changing daily - sometimes hourly - so the "Tactical Plan" comes into play, as the original strategic courses are modified even more to reflect the movement of the ice.

The final fine-tuning is done in the ice-navigator's brain as the ice is observed by eye from the wheelhouse. Minute-by-minute decisions for manoeuvring are based on the interpretation of what the ice-navigator sees through the wheelhouse windows ahead, to the sides and all around. It is this final process which no electronic system has thus far been able to emulate, because the experienced ice-navigator is unconsciously processing in his mind all of the clues presented by the observed ice conditions.

These clues indicate the different types of ice - of which there are seventeen officially recognised ; the selection of dangerous floes from others which do not threaten damage; the onset or release of pressure; the thickness and age of the individual floes; the advance or retreat of the melting processes on the ice-field; the snow-cover concealing dangerous ice features; the effects of wind and tide, and a host of other small details which remain undetected by any electronic sensing systems.

ELECTRONICS CAN BE OF ASSISTANCE TO THE ICE -NAVIGATOR, BUT ARE TOO EXPENSIVE FOR MOST VESSELS

All of the above does not deny that electronics can be of assistance to the ice-navigator, especially where electronic navigation charts are in use. For vessels such as Government icebreakers, which spend their entire lives engaged in ice navigation, there are already some very sophisticated systems which will overlay satellite ice-imagery in real time on radar displays, but these systems are very expensive and out of reach for the vessel which makes occasional forays into ice. Much of the ice-information broadcast by Government Agencies is available in electronic form, sometimes downloaded directly from the Internet, or by subscription to specialised services. The author is personally not aware of any interface where these easily attained ice-maps can be overlaid directly on electronic navigation charts, but sees no reason why it is not technically possible.

It has been the author's observation aboard vessels transiting Arctic and Antarctic waters that the electronic charts for Polar Regions leave a lot to be desired in accuracy of information, as well as the lack of navigational detail. These problems would need to be addressed before ice-chart overlays become standard practice. A more practical idea might be to project the image of an ice-chart directly on to a paper chart, with means to adjust the scale of the projection to match the scale of the chart beneath.

The electronic navigation charts already provide a very rapid means for revision of ship-tracks already programmed (the "Base Plan"), so that a "tactical" modification could be easily generated with an ice-map overlaid on the navigational data. Government ice maps are generated from data collected from various sources (satellite surveillance; aerial reconnaissance; visual observations; ship reports etc.) so they are almost always reporting on ice conditions which existed some hours - or even days - in the past.

The accuracy of forecasts varies enormously, so that tactical ice navigation takes precedence over strategic planning once the vessel has entered the ice. It is therefore very useful to have a rapid means available to make frequent changes to the route, not only to make better progress through the ice, but also to understand the eventual outcome and navigational consequences of frequent modifications.

SENSORS COULD BE VALUABLE IN THE PREVENTION OF SERIOUS ICE-DAMAGE

So far we have been looking at how electronics can assist the ice-navigator in making progress through the ice, but this is not the only contribution that electronics can make to the safe transit of a vessel. For vessels which can expect to spend more time navigating in ice than in open water (i.e: icebreakers and specialised vessels designed for polar work in the oil industry, or bulk carriers exploiting mineral resources in polar regions) it might be worth investing in sensors throughout the ship to indicate to the ice-navigator the punishment inflicted by the ice.

This is particularly necessary in very large vessels, where the wheelhouse may be a long way from the bow, and the force of impacts on the hull will not be physically appreciated on the Bridge. There are many cases of serious ice-damage to the hulls of vessels which have gone completely unnoticed at the moment of impact, only becoming apparent when flooding occurs or an oil-slick appears in the water alongside.

'NAVIGATING BY THE SOLES OF THE FEET' NOT ENOUGH ON LARGER VESSELS

The installation of stress gauges in the fore-part of the hull above, at and below the ice-belt, also along the sides around the waterline, and on the rudder, propellers and shafting, with read-outs and alarms available to the ice-navigator, will go a long way in curbing excessive speed, as the readings delivered by ice-impacts will trigger visual and audible warnings.

On smaller vessels the ice-navigator can - as the saying goes - "navigate by the soles of the feet", being aware of every impact, acceleration and sudden deceleration caused by the ice, which will attract the attention of the Captain if things become violent enough to cause damage. Such is not the case, however, on larger vessels, which must provide feed-back to the ice-navigator for physical sensations which do not penetrate to the command position.

THE PROS AND CONS OF ELECTRONIC REMOTE-SENSING AND SHIP-BORNE SENSING

There has been a lot of research for many years in electronic remote-sensing to determine the different characteristics of ice. So far, the greatest success has been with satellite-borne synthetic aperture radar equipment, and infra-red sensing to observe ice-surface temperature differences, thereby inferring the varying thickness of floes in an ice-field. Laser beams directed downwards from satellites can measure the roughness of the ice, and aircraft fitted with Side Looking Airborne Radar (SLAR) cameras can record the concentration of ice in great detail, even to the extent of identifying floes down to10 metresin diameter.

All of this is wonderful for reporting ice conditions to the mariner for strategic planning, but it is mostly very small scale, requiring special expertise in interpretation of the visual data, thus with limited use in tactical planning for the average seafarer. Research in ship-borne sensing has not been so successful.

The sensor has such a close horizon due to its restricted height above the waterline, dictated by the height of the installation point on the mast or fo'c'sle, that the ice-navigator cannot plan more than the next move ahead, when subsequent moves through the ice must also be considered, just like in a game of chess. The human eye can do a better job for ice this close to the vessel, and the human brain can interpret the situation more quickly.

MARINE RADAR - THE ADVANTAGES AND DISADVANTAGES

Marine radar has been used as a warning system for ice ever since it became commonplace on board ship for navigational purposes. While it gives a good warning that ice is present, it is not a great help in most attempts at finding a safe route through drift ice, because it cannot discriminate between different types of ice. The interpretation of the presentation is complicated by too many radar returns from ice which is not at all dangerous. Radar is thus not a good tool for finding a route through an ice-field. It is, however, very good at showing the limits of ice when approaching from open water. There is a tendency to rely too much on radar to warn of the approach of ice, particularly the most dangerous type of all, which is glacial ice (icebergs) or fragments of them (bergy bits and growlers).

Small bergy bits and growlers often disappear from both radar and visual view in a swell or rough seas, which makes them particularly dangerous. The development of the Automated Radar Plotting Aid (ARPA) and the computerised daylight-view radar screen assist the ice-navigator in ways that the older radars never could. The ability to maintain automatic plots of large numbers of stationary objects (icebergs), azimuth-stabilisation and true motion "north-up" displays give a realistic picture of the physical environment around the vessel, so that the ice-navigator can concentrate on finding a way through the ice by visual observation - still the best approach.

SONAR NOT A SOLUTION

Forward-looking sonar is often advocated for warning of icebergs and their accompanying growlers, but there are very few merchant ships fitted with this, and the chance of locating by sonar small - but dangerous - ice fragments, with sufficient time to avoid collision is quite slim. Large icebergs are generally easily visible to the eye and to radar in darkness or thick weather, so sonar would be a duplication of effort. The only ice of interest to a vessel is at less than10 metresdepth, so a sonar system would have to be set very close to the surface, more likely to be inefficient, especially in rough weather. It would also be very vulnerable to ice damage and the shocks from ice-impact on the stem, probably not worth the cost of installation for the small benefit it might provide.

CONCLUSION -"KEEP LOOKING OUT OF THE WHEELHOUSE WINDOWS"

In conclusion, the art of ice navigation can be aided and enhanced by the use of electronic aids, many more of which are still in development, but the ice-navigator should never rely too much on equipment which can never show the full picture in anything like the detail provided by a good pair of eyes - with binoculars for the distant stuff. Remember the mantra, "keep looking out of the wheelhouse windows" and you should not come to too much harm.

IMPORTANT POINTS FOR ICE NAVIGATION OF SHIPS

Navigating in ice waters can be a real task for ships, as the later moves cracking and smashing through the frozen and frigid seas. While moving towards subzero temperature with ice covered waters, the ship's captain has to be extremely cautious and must pay utmost attention to the type of ice, thickness, and its exact location in the subzero navigation areas. If there's any kind of misjudgment during ice navigation, a detour from the navigable route would lead to additional fuel wastage and might also get the vessel stuck in thick ice leading to dangerous situation and damage.

The existence of ice on seawater corresponds to a major restraint for ships and offshore operations at high latitudes in both the hemispheres. The sea-ice, which on an average is 2-3 m thick, can be pierced only by specially designed ice-strengthened vessels or icebreakers with an appropriate ice class. Most merchant ships and fishing vessels which are not ice-strengthened must consequently keep away from all ice waters and sub-temperature areas. In many places, where the concentration of ice is maximum and the ice pressure is highest, even the most powerful of the icebreakers have problems with ice navigation.

To avoid such mishaps during ice navigation, an ice pilot and ice breakers are normally provided for commercial vessel operations which help to safely guide the vessels through the ice field to and from their destinations. In addition to the ice pilot being on board and the ice-breaker's relentless assistance, a few other measures must be kept in mind by the vessel's crew during the ice navigation.

MANOEUVRING IN ICE

First of all, it is imperative to understand that if any alternative route is available for the ship, ice water should be avoided at all costs. However, if ice navigation is inevitable, it should be made at right angles to the leeward edge where the ice is loose or broken. While manoeuvring through ice if a floe cannot be avoided then it should be hit squarely with the stem. Note that a glancing blow may damage the ship's shell plating or throw the vessel off course causing another unavoidable blow.

Entry in ice should always be done at low speeds to avoid any sort of damage. Once into the pack, the vessel's speed can be increased so as to maintain headway and control so as to never lose all way off and avoid the ice floes to close in on the hull, rudder and propeller. If the ship is stopped by heavy concentration of ice the rudder should be put amidships and the engines should be kept turning slowly ahead. This will wash away the ice that is accumulated astern and will help the vessel to fall back. In a close pack during ice navigation, avoid sharp alterations of course and keep the speed enough for steerage way. Full rudder movements should be avoided or used only in cases of emergencies.

LOOKOUT

Always keep vigilant lookout for leads (navigable channel within an ice field) through ice. Additional lookouts should be posted forward or at higher ends for safety concerns. Conning should be carried out from the ship's bridge to get a better view of the ice accumulation. Keep in mind that at all times the stern must be observed for rudders' movement so as to avoid a floe from actually moving the stern towards it. In such cases, it is advised to post men right aft with torches, whistles, walkie-talkies, etc. to make sure that the bridge is informed immediately in case the propeller is in any kind of danger. This is extremely important in twin screw vessels. Reduce speed if the ice goes under the ship.

ENGINE CARE

During ice navigation, engines should be kept running at all times and under maneuvering conditions in such a way that the ahead and astern movements can be easily carried out without time delay. Similarly, engine movements from ahead to astern and vice-verse should be made cautiously to avoid stressing the engine mechanisms in low temperatures, which could be unfavorable to the ship's engine parts. Also, when ice approaches the stern of the vessel while maneuvering bursts of the engines should be given accordingly to keep ice from accumulating.

NAVIGATION AT NIGHT

As far as possible, avoid navigating through ice at night. It is preferred to "heave to" since the leads or lanes cannot be seen. Most ice navigators stop the vessel along the edge of the ice and leave the vessel drifting along with the pack. At nights, seawater lubricated tail end shafts are in the danger of getting frozen. To avoid from freezing, vessels with single screws should have their aft peak tank filled with water and have it kept warm by means of steam hose injection, or other alternative means. The vessel should keep her engines running with propeller on low RPM so as to avoid seizure by ice.

ANCHORING

Anchoring in heavy concentrations of ice should be avoided; if ice is moving then its force may break the cable. When conditions permit, anchoring can be carried out and it must be done in light brash ice, rotten ice or widely scattered floes with the main engine on immediate notice. Anchor should be brought in as soon as the wind threatens to move ice onto the vessel.

Even with the advent of new techniques and technologies for ice navigation such as radar sensor images through cloud cover, infra-red images, and satellite images for a larger view of the surroundings around the vessel, it is vital to understand that ship's operations of any sort under the influence of sea-ice are not only dangerous but also life threatening, and utmost care must be taken while navigating through such ice areas.

10

Ocean Current and Tide

An ocean current is a continuous, directed movement of ocean water generated by the forces acting upon this mean flow, such as breaking waves, wind, Coriolis effect, cabbeling, temperature and salinity differences and tides caused by the gravitational pull of the Moon and the Sun. Depth contours, shoreline configurations and interaction with other currents influence a current's direction and strength. A deep current is any ocean current at a depth of greater than 100m. A part of oceanography is the science studying ocean currents. Ocean currents can flow for great distances, and together they create the great flow of the global conveyor belt which plays a dominant part in determining the climate of many of the Earth's regions. Perhaps the most striking example is the Gulf Stream, which makes northwest Europe much more temperate than any other region at the same latitude. Another example is Lima, Peru, where the climate is cooler (sub-tropical) than the tropical latitudes in which the area is located, due to the effect of the Humboldt Current.

FUNCTION

Surface ocean currents are generally wind-driven and develop their typical clockwise spirals in the northern hemisphere and counter-clockwise rotation in the southern hemisphere because of the imposed wind stresses. In wind-driven current, the Ekman spiral effect results in the currents flowing at an angle to the driving winds. The areas of surface ocean currents move somewhat with the seasons; this is most notable in equatorial currents. Ocean basins generally have a non-symmetric surface current, in that the eastern equatorward-flowing branch is broad and diffuse whereas the western poleward-flowing branch is very narrow. These western boundary currents (of which the gulf stream is an example) are a consequence of basic fluid dynamics.

Deep ocean currents are driven by density and temperature gradients. Thermohaline circulation, also known as the ocean's conveyor belt, refers to the deep ocean density-driven ocean basin currents. These currents, which flow under the surface of the ocean and are thus hidden from immediate detection, are called submarine rivers. These are currently being researched using a fleet of underwater robots called Argo. Upwelling and downwelling areas in the oceans are areas where significant vertical movement of ocean water is observed. Surface currents make up about 8% of all the water in the ocean. Surface currents are generally restricted to the upper 400 m (1,300 ft) of the ocean. The movement of deep water in the ocean basins is by density driven forces and gravity. The density difference is a function of different temperatures and salinity. Deep waters sink into the deep ocean basins at high latitudes where the temperatures are cold enough to cause the density to increase. Ocean currents are measured in Sverdrup (Sv), where 1Sv is equivalent to a volume flow rate of 1,000,000 m (35,000,000 cu ft) per second.

IMPORTANCE

Knowledge of surface ocean currents is essential in reducing costs of shipping, since traveling with them reduces fuel costs. In the sail-ship era knowledge was even more essential. A good example of this is the Agulhas current, which long prevented Portuguese sailors from reaching India. Even today, the round-the-world sailing competitors employ surface currents to their benefit. Ocean currents are also very important in the dispersal of many life forms. An example is the life-cycle of the European Eel. Ocean currents are important in the study of marine debris, and vice versa. These currents also affect temperatures throughout the world. For example, the current that brings warm water up the north Atlantic to northwest Europe stops ice from forming by the shores, which would block ships from entering and exiting ports.

OSCAR: NEAR-REALTIME GLOBAL OCEAN SURFACE CURRENT DATA SET

The OSCAR near-realtime global ocean circulation data set is based on NOAA and NASA satellite data (sea level altimetry, surface vector winds, and SST). The data set extends from 1993-present and is available at 1-degree and 1/3-degree resolution. The OSCAR data are continuously updated on an interactive website from which users can create customized graphics and download the data. A section of the website provides validation studies in the form of graphics comparing OSCAR data with moored buoys and global drifters.

OSCAR data are used extensively in climate studies. Monthly maps and anomalies have been published in the monthly Climate Diagnostic Bulletin since 2001, and are used routinely to monitor ENSO and to test prediction models. OSCAR currents are routinely used to evaluate the surface currents in Global Circulation Models (GCMs), for example in NCEP Global Ocean Data Assimilation System (GODAS) and European Centre for Medium-Range Weather Forecasts (ECMWF).

INFRAGRAVITY WAVE

Infragravity waves are surface gravity waves with frequencies lower than the wind waves - consisting of both wind sea and swell - so corresponding with the part of the wave spectrum lower than the frequencies directly generated by forcing through the wind.

Infragravity waves consist, among others, of long-period oceanic waves generated along continental coastlines by nonlinear wave interactions of storm-forced shoreward-propagating ocean swells. These differ from normal oceanic gravity waves, which are created by wind acting on the surface of the sea. Normal gravity waves typically have a frequency on the order of 50 millihertz (i.e., a period of 20 seconds). Interactions of these waves with coastlines filters out the frequencies with periods about 30 seconds, but nonlinear processes convert some of this energy to subharmonics with periods ranging from 50 seconds (20 mHz) to 350 seconds (3 mHz). Infragravity waves are these subharmonics of the impinging gravity waves.

Technically infragravity waves are simply a subcategory of gravity waves and refer to all gravity waves with periods greater than 30 s. Although they include phenomena such as tides and oceanic Rossby waves, in the common literature their use is limited to gravity waves that are generated by the topography of the bottom. The term "infragravity wave" appears to have been coined by Walter Munk in 1950.

GENERATION

As a result of geology, infragravity-wave-induced large-scale bedforms (e.g., bars), and biologic process (e.g., reefs) the shoreline and the near-shore features of the sea floor often has a periodic character. Coastal sand bars are a significant contributor to the generation of infragravity waves and are shaped by them. On the inner side of a sand bar, the size of the bar is determined by the length of short wavelength wind-generated gravity waves. On the outer side of the bar the bar shape is dictated by the length of the infragravity waves which correlate to and are driven by the groups of short waves. Similarly, coral reefs are effective in generating infragravity waves; in the case of coral reefs, the infragravity periods are established by resonances with the reef itself.

IMPACT

Infragravity waves generated along the Pacific coast of North America have been observed to propagate transoceanically to Antarctica and there to impinge on the Ross Ice Shelf. Their frequencies more closely couple with the ice shelf natural frequencies and they produce a larger amplitude ice shelf movement than the normal ocean swell of gravity waves. Further, they are not damped by sea ice as normal ocean swell is. As a result they flex floating ice shelves such as the Ross Ice Shelf; this flexure contributes significantly to the breakup on the ice shelf.

EQUATORIAL WAVES

Equatorial waves are ocean waves trapped close to the equator, meaning that they decay rapidly away from the equator, but can propagate in the longitudinal and vertical directions. Wave trapping is the result of the Earth's rotation and its spherical shape which combine to cause the magnitude of the Coriolis force to increase rapidly away from the equator. Equatorial waves are present in both the tropical atmosphere and ocean and play an important role in the evolution of many climate phenomena such as El Niño. Many physical processes may excite equatorial waves including, in the case of the atmosphere, diabatic heat release associated with cloud formation, and in the case of the ocean, anomalous changes in the strength or direction of the trade winds. Equatorial waves may be separated into a series of subclasses depending on their fundamental dynamics (which also influences their typical periods and speeds and directions of propagation). At shortest periods are the equatorial gravity waves while the longest periods are associated with the equatorial Rossby waves. In addition to these two extreme subclasses, there are two special subclasses of equatorial waves known as the mixed Rossby-gravity wave (also known as the Yanai wave) and the equatorial Kelvin wave. The latter two share the characteristics that they can have any period and also that they may carry energy only in an eastward (never westward) direction. The remainder of this article discusses the relationship between the period of these waves, their wavelength in the zonal direction and their speeds for a simplified ocean.

EQUATORIAL KELVIN WAVES

Discovered by Lord Kelvin, coastal Kelvin waves are trapped close to coasts and propagate along coasts in the Northern Hemisphere such that the coast is to the right of the alongshore direction of propagation (and to the left in the Southern Hemisphere). Equatorial Kelvin waves behave somewhat as if there were a wall at the equator - so that the equator is to the right of the direction of along-equator propagation in the Northern Hemisphere and to the left of the direction of propagation in the Southern Hemisphere, both of which are consistent with eastward propagation along the equator. The governing equations for these equatorial waves are similar to those presented above, except that there is no meridional velocity component (that is, no flow in the north-south direction).

- the continuity equation (accounting for the effects of horizontal convergence and divergence):

$$\frac{\partial \eta}{\partial t} + c^2 \frac{\partial u}{\partial x} = 0$$

- the u-momentum equation (zonal wind component):

$$\frac{\partial u}{\partial t} = -\frac{\partial \eta}{\partial x}$$

- the v-momentum equation (meridional wind component):

$$uey @ 0\frac{»i}{»y}.$$

The solution to these equations yields the following phase speed: $c = gH$; this result is the same speed as for shallow-water gravity waves without the effect of Earth's rotation. Therefore, these waves are non-dispersive (because the phase speed is not a function of the zonal wavenumber). Also, these Kelvin waves only propagate towards the east (because as ? approaches zero, y approaches infinity).

CONNECTION TO EL NINO SOUTHERN OSCILLATION

Kelvin waves have been connected to El Nino (beginning in the Northern Hemisphere winter months) in recent years in terms of precursors to this atmospheric and oceanic phenomenon. Many scientists have utilized coupled atmosphere-ocean models to simulate an El Nino Southern Oscillation (ENSO) event and have stated that the Madden-Julian oscillation (MJO) can trigger oceanic Kelvin waves throughout its 30-60 day cycle or the latent heat of condensation can be released (from intense convection) resulting in Kelvin waves as well; this process can then signal the onset of an El Nino event. The weak low pressure in the Indian Ocean (due to the MJO) typically propagates eastward into the North Pacific Ocean and can produce easterly winds.

These easterly winds can transfer West Pacific warm water toward the east, thereby exciting a Kelvin wave, which in this sense can be thought of as a warm-water anomaly that travels under the ocean's surface at a depth of about 150 meters. This wave can be observed at the surface by a slight rise in sea surface height of about 8 cm (associated with a depression of the thermocline) and an SST increase that covers hundreds of square miles across the surface of the ocean. If the Kelvin wave hits the South American coast (specifically Ecuador), its warm water gets transferred upward, which creates a large warm pool at the surface. That warm water also starts to flow southward along the coast of Peru and north towards Central America and Mexico, and may reach parts of Northern California; the wave can then be tracked primarily using an array of 70 buoys anchored along the entire width of equatorial Pacific Ocean, from Papua New Guinea to the Ecuador coast. Temperature sensors are placed at different depths along the buoys' anchor-lines and are then able to record sub-surface water temperature.

The sensors send their data in real-time using a satellite to a central processing facility. These temperature measurements are then compared and contrasted to historically- and seasonally-adjusted average water temperatures for each buoy location. Some results indicate deviations from the 'normal' expected temperatures. Such deviations are referred to as anomalies and can be thought of as either warmer-than-normal (El Nino) or cooler-than-normal (La Nina) conditions. The overall ENSO cycle can be explained as follows (in terms of the wave propagation throughout the Pacific Ocean): ENSO begins with a warm pool traveling from the western Pacific to the eastern Pacific in the form of Kelvin waves (the waves carry the warm SSTs) that resulted from the MJO. After approximately 3 to 4 months of propagation across the Pacific (along the equatorial region), the Kelvin waves reach the western coast of South America and interact (merge/mix) with the cooler Peru current system.

This causes a rise in sea levels and sea level temperatures in the general region. Upon reaching the coast, the water turns to the north and south and results in El Nino conditions to the south. Because of the changes in sea-level and sea-temperature due to the Kelvin waves, an infinite number of Rossby waves are generated and move back over the Pacific. Rossby waves then enter the equation and, as previously stated, move at lower velocities than the Kelvin waves and can take anywhere from nine months to four years to fully cross the Pacific Ocean basin (from boundary to boundary). And because these waves are equatorial in nature, they decay rapidly as distance from the equator increases; thus, as they move away from the equator, their speed decreases as well, resulting in a wave delay.

When the Rossby waves reach the western Pacific they ricochet off the coast and become Kelvin waves and then propagate back across the Pacific in the direction of the South America coast. Upon return, however, the waves decrease the sea-level (reducing the depression in the thermocline) and sea surface temperature, thereby returning the area to normal or sometimes La Nina conditions. In terms of climate modeling and upon coupling the atmosphere and the ocean, an ENSO model typically contains the following dynamical equations:

- 3 primitive equations for the atmosphere (as mentioned above) with the inclusion of frictional parameterizations: 1) u-momentum equation, 2) v-momentum equation, and 3) continuity equation
- 4 primitive equations for the ocean (as stated below) with the inclusion of frictional parameterizations:
- – u-momentum,

$$\frac{\partial u}{\partial t} - v\beta y = \frac{\tau_x}{\rho h},$$

- – v-momentum,

$$\frac{\partial v}{\partial t} - u\beta y = \frac{\tau_y}{\rho h},$$

- – continuity,

$$\frac{\partial h}{\partial t} + h\left(\frac{\partial u}{\partial x} + \frac{\partial v}{\partial y}\right) - K_E T = 0,$$

- – thermodynamic energy,

$$\frac{»T}{»t} \;.\; u\frac{»T}{»x} \;0\; K_T h \;@\; 0.$$

Note that h is the depth of the fluid (similar to the equivalent depth and analoguous to H in the primitive equations listed above for Rossby-gravity and Kelvin waves), KT is temperature diffusion, KE is eddy diffusivity, and w is the wind stress in either the x or y directions.

11

Tide

Tides are the rise and fall of sea levels caused by the combined effects of the gravitational forces exerted by the Moon and the Sun and the rotation of the Earth. Some shorelines experience two almost equal high tides and two low tides each day, called a semi-diurnal tide. Some locations experience only one high and one low tide each day, called a diurnal tide. Some locations experience two uneven tides a day, or sometimes one high and one low each day; this is called a mixed tide.

The times and amplitude of the tides at a locale are influenced by the alignment of the Sun and Moon, by the pattern of tides in the deep ocean, by the amphidromic systems of the oceans, and by the shape of the coastline and near-shore bathymetry.

Tides vary on timescales ranging from hours to years due to numerous influences. To make accurate records, tide gauges at fixed stations measure the water level over time. Gauges ignore variations caused by waves with periods shorter than minutes. These data are compared to the reference (or datum) level usually called mean sea level.

While tides are usually the largest source of short-term sea-level fluctuations, sea levels are also subject to forces such as wind and barometric pressure changes, resulting in storm surges, especially in shallow seas and near coasts. Tidal phenomena are not limited to the oceans, but can occur in other systems whenever a gravitational field that varies in time and space is present. For example, the solid part of the Earth is affected by tides, though this is not as easily seen as the water tidal movements.

CHARACTERISTICS

Tide changes proceed via the following stages:

- Sea level rises over several hours, covering the intertidal zone; flood tide.
- The water rises to its highest level, reaching high tide.
- Sea level falls over several hours, revealing the intertidal zone; ebb tide.
- The water stops falling, reaching low tide.

Tides produce oscillating currents known as tidal streams. The moment that the tidal current ceases is called slack water or slack tide. The tide then reverses direction and is said to be turning. Slack water usually occurs near high water and low water. But there are locations where the moments of slack tide differ significantly from those of high and low water.

Tides are most commonly semi-diurnal (two high waters and two low waters each day), or diurnal (one tidal cycle per day). The two high waters on a given day are typically not the same height (the daily inequality); these are the higher high water and the lower high water in tide tables. Similarly, the two low waters each day are the higher low water and the lower low water. The daily inequality is not consistent and is generally small when the Moon is over the equator.

TIDAL CONSTITUENTS

Tidal changes are the net result of multiple influences that act over varying periods. These influences are called tidal constituents. The primary constituents are the Earth's rotation, the positions of the Moon and the Sun relative to Earth, the Moon's altitude (elevation) above the Earth's equator, and bathymetry. Variations with periods of less than half a day are called harmonic constituents. Conversely, cycles of days, months, or years are referred to as long period constituents.

The tidal forces affect the entire earth, but the movement of the solid Earth is only centimeters. The atmosphere is much more fluid and compressible so its surface moves kilometers, in the sense of the contour level of a particular low pressure in the outer atmosphere.

PRINCIPAL LUNAR SEMI-DIURNAL CONSTITUENT

In most locations, the largest constituent is the "principal lunar semi-diurnal", also known as the M2 (or M2) tidal constituent. Its period is about 12 hours and 25.2 minutes, exactly half a tidal lunar day, which is the average time separating one lunar zenith from the next, and thus is the time required for the Earth to rotate once relative to the Moon. Simple tide clocks track this constituent. The lunar day is longer than the Earth day because the Moon orbits in the same direction the Earth spins. This is analogous to the minute hand on a watch crossing the hour hand at 12:00 and then again at about 1:05½ (not at 1:00).

The Moon orbits the Earth in the same direction as the Earth rotates on its axis, so it takes slightly more than a day-about 24 hours and 50 minutes-for the Moon to return to the same location in the sky. During this time, it has passed overhead (culmination) once and underfoot once (at an hour angle of 00:00 and 12:00 respectively), so in many places the period of strongest tidal forcing is the above mentioned, about 12 hours and 25 minutes. The moment of highest tide is not necessarily when the Moon is nearest to zenith or nadir, but the period of the forcing still determines the time between high tides. Because the gravitational field created by the Moon weakens with distance from the Moon, it exerts a slightly stronger than average force on the side of the Earth facing the Moon, and a slightly weaker force on the opposite side. The Moon thus tends to "stretch" the Earth slightly along the line connecting the two bodies.

The solid Earth deforms a bit, but ocean water, being fluid, is free to move much more in response to the tidal force, particularly horizontally. As the Earth rotates, the magnitude and direction of the tidal force at any particular point on the Earth's surface change constantly; although the ocean never reaches equilibrium-there is never time for the fluid to "catch up" to the state it would eventually reach if the tidal force were constant-the changing tidal force nonetheless causes rhythmic changes in sea surface height.

SEMI-DIURNAL RANGE DIFFERENCES

When there are two high tides each day with different heights (and two low tides also of different heights), the pattern is called a mixed semi-diurnal tide.

RANGE VARIATION: SPRINGS AND NEAPS

The semi-diurnal range (the difference in height between high and low waters over about half a day) varies in a two-week cycle. Approximately twice a month, around new moon and full moon when the Sun, Moon, and Earth form a line (a condition known as syzygy), the tidal force due to the sun reinforces that due to the Moon. The tide's range is then at its maximum; this is called the spring tide, or just springs. It is not named after the season, but, like that word, derives from the meaning "jump, burst forth, rise", as in a natural spring.

When the Moon is at first quarter or third quarter, the sun and Moon are separated by 90° when viewed from the Earth, and the solar tidal force partially cancels the Moon's. At these points in the lunar cycle, the tide's range is at its minimum; this is called the neap tide, or neaps (a word of uncertain origin). Spring tides result in high waters that are higher than average, low waters that are lower than average, 'slack water' time that is shorter than average, and stronger tidal currents than average. Neaps result in less-extreme tidal conditions. There is about a seven-day interval between springs and neaps.

LUNAR ALTITUDE

The changing distance separating the Moon and Earth also affects tide heights. When the Moon is closest, at perigee, the range increases, and when it is at apogee, the range shrinks. Every 7½ lunations (the full cycles from full moon to new to full), perigee coincides with either a new or full moon causing perigean spring tides with the largest tidal range. Even at its most powerful this force is still weak causing tidal differences of inches at most.

BATHYMETRY

The shape of the shoreline and the ocean floor changes the way that tides propagate, so there is no simple, general rule that predicts the time of high water from the Moon's position in the sky. Coastal characteristics such as underwater bathymetry and coastline shape mean that individual location characteristics affect tide forecasting; actual high water time and height may differ from model predictions due to the coastal morphology's effects on tidal flow. However, for a given location the relationship between lunar altitude and the time of high or low tide (the lunitidal interval) is relatively constant and predictable, as is the time of high or low tide relative to other points on the same coast. For example, the high tide at Norfolk, Virginia, predictably occurs approximately two and a half hours before the Moon passes directly overhead.

Land masses and ocean basins act as barriers against water moving freely around the globe, and their varied shapes and sizes affect the size of tidal frequencies. As a result, tidal patterns vary. For example, in the U.S., the East coast has predominantly semi-diurnal tides, as do Europe's Atlantic coasts, while the West coast predominantly has mixed tides.

OTHER CONSTITUENTS

These include solar gravitational effects, the obliquity (tilt) of the Earth's equator and rotational axis, the inclination of the plane of the lunar orbit and the elliptical shape of the Earth's orbit of the sun. A compound tide (or overtide) results from the shallow-water interaction of its two parent waves.

PHASE AND AMPLITUDE

Because the M2 tidal constituent dominates in most locations, the stage or phase of a tide, denoted by the time in hours after high water, is a useful concept. Tidal stage is also measured in degrees, with 360° per tidal cycle. Lines of constant tidal phase are called cotidal lines, which are analogous to contour lines of constant altitude on topographical maps. High water is reached simultaneously along the cotidal lines extending from the coast out into the ocean, and cotidal lines (and hence tidal phases) advance along the coast. Semi-diurnal and long phase constituents are measured from high water, diurnal from maximum flood tide. This and the discussion that follows is precisely true only for a single tidal constituent.

For an ocean in the shape of a circular basin enclosed by a coastline, the cotidal lines point radially inward and must eventually meet at a common point, the amphidromic point. The amphidromic point is at once cotidal with high and low waters, which is satisfied by zero tidal motion. (The rare exception occurs when the tide encircles an island, as it does around New Zealand, Iceland and Madagascar.) Tidal motion generally lessens moving away from continental coasts, so that crossing the cotidal lines are contours of constant amplitude (half the distance between high and low water) which decrease to zero at the amphidromic point. For a semi-diurnal tide the amphidromic point can be thought of roughly like the center of a clock face, with the hour hand pointing in the direction of the high water cotidal line, which is directly opposite the low water cotidal line. High water rotates about the amphidromic point once every 12 hours in the direction of rising cotidal lines, and away from ebbing cotidal lines. This rotation is generally clockwise in the southern hemisphere and counterclockwise in the northern hemisphere, and is caused by the Coriolis effect. The difference of cotidal phase from the phase of a reference tide is the epoch. The reference tide is the hypothetical constituent "equilibrium tide" on a landless Earth measured at 0° longitude, the Greenwich meridian.

In the North Atlantic, because the cotidal lines circulate counterclockwise around the amphidromic point, the high tide passes New York Harbor approximately an hour ahead of Norfolk Harbor. South of Cape Hatteras the tidal forces are more complex, and cannot be predicted reliably based on the North Atlantic cotidal lines.

PHYSICS

HISTORY OF TIDAL PHYSICS

Investigation into tidal physics was important in the early development of heliocentrism and celestial mechanics, with the existence of two daily tides being explained by the Moon's gravity. Later the daily tides were explained more precisely by the interaction of the Moon's and the sun's gravity. Galileo Galilei in his 1632 Dialogue Concerning the Two Chief World Systems, whose working title was Dialogue on the Tides, gave an explanation of the tides. The resulting theory, however, was incorrect as he attributed the tides to the sloshing of water caused by the Earth's movement around the sun. He hoped to provide mechanical proof of the Earth's movement - the value of his tidal theory is disputed. At the same time Johannes Kepler correctly suggested that the Moon caused the tides, which he based upon ancient observations and correlations, an explanation which was rejected by Galileo. It was originally mentioned in Ptolemy's Tetrabiblos as having derived from ancient observation.

Isaac Newton (1642-1727) was the first person to explain tides as the product of the gravitational attraction of astronomical masses. His explanation of the tides (and many other phenomena) was published in the Principia (1687) and used his theory of universal gravitation to explain the lunar and solar attractions as the origin of the tide-generating forces. Newton and others before

Pierre-Simon Laplace worked the problem from the perspective of a static system (equilibrium theory), that provided an approximation that described the tides that would occur in a non-inertial ocean evenly covering the whole Earth. The tide-generating force (or its corresponding potential) is still relevant to tidal theory, but as an intermediate quantity (forcing function) rather than as a final result; theory must also consider the Earth's accumulated dynamic tidal response to the applied forces, which response is influenced by bathymetry, Earth's rotation, and other factors.

In 1740, the Académie Royale des Sciences in Paris offered a prize for the best theoretical essay on tides. Daniel Bernoulli, Leonhard Euler, Colin Maclaurin and Antoine Cavalleri shared the prize.

Maclaurin used Newton's theory to show that a smooth sphere covered by a sufficiently deep ocean under the tidal force of a single deforming body is a prolate spheroid (essentially a three dimensional oval) with major axis directed toward the deforming body. Maclaurin was the first to write about the Earth's rotational effects on motion. Euler realized that the tidal force's horizontal component (more than the vertical) drives the tide. In 1744 Jean le Rond d'Alembert studied tidal equations for the atmosphere which did not include rotation.

Pierre-Simon Laplace formulated a system of partial differential equations relating the ocean's horizontal flow to its surface height, the first major dynamic theory for water tides. The Laplace tidal equations are still in use today. William Thomson, 1st Baron Kelvin, rewrote Laplace's equations in terms of vorticity which allowed for solutions describing tidally driven coastally trapped waves, known as Kelvin waves. Others including Kelvin and Henri Poincaré further developed Laplace's theory. Based on these developments and the lunar theory of E W Brown describing the motions of the Moon, Arthur Thomas Doodson developed and published in 1921 the first modern development of the tide-generating potential in harmonic form: Doodson distinguished 388 tidal frequencies. Some of his methods remain in use.

FORCES

The tidal force produced by a massive object (Moon, hereafter) on a small particle located on or in an extensive body (Earth, hereafter) is the vector difference between the gravitational force exerted by the Moon on the particle, and the gravitational force that would be exerted on the particle if it were located at the Earth's center of mass. The solar gravitational force on the Earth is on average 179 times stronger than the lunar, but because the Sun is on average 389 times farther from the Earth, its field gradient is weaker. The solar tidal force is 46% as large as the lunar. More precisely, the lunar tidal acceleration (along the Moon-Earth axis, at the Earth's surface) is about 1.1 × 10 g, while the solar tidal acceleration (along the Sun-Earth axis, at the Earth's surface) is about 0.52 × 10 g, where g is the gravitational acceleration at the Earth's surface. Venus has the largest effect of the other planets, at 0.000113 times the solar effect. The ocean's surface is closely approximated by an equipotential surface, (ignoring ocean currents) commonly referred to as the geoid. Since the gravitational force is equal to the potential's gradient, there are no tangential forces on such a surface, and the ocean surface is thus in gravitational equilibrium. Now consider the effect of massive external bodies such as the Moon and Sun.

These bodies have strong gravitational fields that diminish with distance in space and which act to alter the shape of an equipotential surface on the Earth. This deformation has a fixed spatial orientation relative to the influencing body. The Earth's rotation relative to this shape causes the daily tidal cycle. Gravitational forces follow an inverse-square law (force is inversely proportional to the square of the distance), but tidal forces are inversely proportional to the cube of the distance. The ocean surface moves because of the changing tidal equipotential, rising when the tidal potential is high, which occurs on the parts of the Earth nearest to and furthest from the Moon. When the tidal equipotential changes, the ocean surface is no longer aligned with it, so the apparent direction of the vertical shifts. The surface then experiences a down slope, in the direction that the equipotential has risen.

LAPLACE'S TIDAL EQUATIONS

Ocean depths are much smaller than their horizontal extent. Thus, the response to tidal forcing can be modelled using the Laplace tidal equations which incorporate the following features:

1. The vertical (or radial) velocity is negligible, and there is no vertical shear-this is a sheet flow.
2. The forcing is only horizontal (tangential).
3. The Coriolis effect appears as an inertial force (fictitious) acting laterally to the direction of flow and proportional to velocity.
4. The surface height's rate of change is proportional to the negative divergence of velocity multiplied by the depth. As the horizontal velocity stretches or compresses the ocean as a sheet, the volume thins or thickens, respectively.

The boundary conditions dictate no flow across the coastline and free slip at the bottom. The Coriolis effect (inertial force) steers currents moving towards the equator to the west and toward the east for flows moving away from the equator, allowing coastally trapped waves. Finally, a dissipation term can be added which is an analog to viscosity.

AMPLITUDE AND CYCLE TIME

The theoretical amplitude of oceanic tides caused by the moon is about 54 centimetres (21 in) at the highest point, which corresponds to the amplitude that would be reached if the ocean possessed a uniform depth, there were no landmasses, and the Earth were rotating in step with the moon's orbit. The sun similarly causes tides, of which the theoretical amplitude is about 25 centimetres (9.8 in) (46% of that of the moon) with a cycle time of 12 hours. At spring tide the two effects add to each other to a theoretical level of 79 centimetres (31 in), while at neap tide the theoretical level is reduced to 29 centimetres (11 in).

Since the orbits of the Earth about the sun, and the moon about the Earth, are elliptical, tidal amplitudes change somewhat as a result of the varying Earth-sun and Earth-moon distances. This causes a variation in the tidal force and theoretical amplitude of about ±18% for the moon and ±5% for the sun. If both the sun and moon were at their closest positions and aligned at new moon, the theoretical amplitude would reach 93 centimetres (37 in). Real amplitudes differ considerably, not only because of depth variations and continental obstacles, but also because wave propagation across the ocean has a natural period of the same order of magnitude as the rotation period: if there were no land masses, it would take about 30 hours for a long wavelength surface wave to propagate along the equator halfway around the Earth (by comparison, the Earth's lithosphere has a natural period of about 57 minutes). Earth tides, which raise and lower the bottom of the ocean, and the tide's own gravitational self attraction are both significant and further complicate the ocean's response to tidal forces.

DISSIPATION

Earth's tidal oscillations introduce dissipation at an average rate of about 3.75 terawatt. About 98% of this dissipation is by marine tidal movement. Dissipation arises as basin-scale tidal flows drive smaller-scale flows which experience turbulent dissipation. This tidal drag creates torque on the moon that gradually transfers angular momentum to its orbit, and a gradual increase in Earth-moon separation.

The equal and opposite torque on the Earth correspondingly decreases its rotational velocity. Thus, over geologic time, the moon recedes from the Earth, at about 3.8 centimetres (1.5 in)/year, lengthening the terrestrial day. Day length has increased by about 2 hours in the last 600 million years. Assuming (as a crude approximation) that the deceleration rate has been constant, this would imply that 70 million years ago, day length was on the order of 1% shorter with about 4 more days per year.

OBSERVATION AND PREDICTION

HISTORY

From ancient times, tidal observation and discussion has increased in sophistication, first marking the daily recurrence, then tides' relationship to the sun and moon. Pytheas travelled to the British Isles about 325 BC and seems to be the first to have related spring tides to the phase of the moon. In the 2nd century BC, the Babylonian astronomer, Seleucus of Seleucia, correctly described the phenomenon of tides in order to support his heliocentric theory. He correctly theorized that tides were caused by the moon, although he believed that the interaction was mediated by the pneuma. He noted that tides varied in time and strength in different parts of the world. According to Strabo (1.1.9), Seleucus was the first to link tides to the lunar attraction, and that the height of the tides depends on the moon's position relative to the sun.

The Naturalis Historia of Pliny the Elder collates many tidal observations, e.g., the spring tides are a few days after (or before) new and full moon and are highest around the equinoxes, though Pliny noted many relationships now regarded as fanciful. In his Geography, Strabo described tides in the Persian Gulf having their greatest range when the moon was furthest from the plane of the equator. All this despite the relatively small amplitude of Mediterranean basin tides. (The strong currents through the Euripus Strait and the Strait of Messina puzzled Aristotle.) Philostratus discussed tides in Book Five of The Life of Apollonius of Tyana. Philostratus mentions the moon, but attributes tides to "spirits". In Europe around 730 AD, the Venerable Bede described how the rising tide on one coast of the British Isles coincided with the fall on the other and described the time progression of high water along the Northumbrian coast.

The first tide table in China was recorded in 1056 AD primarily for visitors wishing to see the famous tidal bore in the Qiantang River. The first known British tide table is thought to be that of John Wallingford, who died Abbot of St. Albans in 1213, based on high water occurring 48 minutes later each day, and three hours earlier at the Thames mouth than upriver at London. William Thomson (Lord Kelvin) led the first systematic harmonic analysis of tidal records starting in 1867. The main result was the building of a tide-predicting machine using a system of pulleys to add together six harmonic time functions. It was "programmed" by resetting gears and chains to adjust phasing and amplitudes. Similar machines were used until the 1960s.

The first known sea-level record of an entire spring-neap cycle was made in 1831 on the Navy Dock in the Thames Estuary. Many large ports had automatic tide gage stations by 1850. William Whewell first mapped co-tidal lines ending with a nearly global chart in 1836. In order to make these maps consistent, he hypothesized the existence of amphidromes where co-tidal lines meet in the mid-ocean. These points of no tide were confirmed by measurement in 1840 by Captain Hewett, RN, from careful soundings in the North Sea.

TIMING

The tidal forces due to the Moon and Sun generate very long waves which travel all around the ocean following the paths shown in co-tidal charts. The time when the crest of the wave reaches a port then gives the time of high water at the port. The time taken for the wave to travel around the ocean also means that there is a delay between the phases the moon and their effect on the tide. Springs and neaps in the North Sea, for example, are two days behind the new/full moon and first/third quarter moon. This is called the tide's age.

The ocean bathymetry greatly influences the tide's exact time and height at a particular coastal point. There are some extreme cases; the Bay of Fundy, on the east coast of Canada, is often stated to have the world's highest tides because of its shape, bathymetry, and its distance from the continental shelf edge. Measurements made in November 1998 at Burntcoat Head in the Bay of Fundy recorded a maximum range of 16.3 metres (53 ft) and a highest predicted extreme of 17 metres (56 ft). Similar measurements made in March 2002 at Leaf Basin, Ungava Bay in northern Quebec gave similar values (allowing for measurement errors), a maximum range of 16.2 metres (53 ft) and a highest predicted extreme of 16.8 metres (55 ft). Ungava Bay and the Bay of Fundy lie similar distances from the continental shelf edge, but Ungava Bay is free of pack ice for only about four months every year while the Bay of Fundy rarely freezes.

Southampton in the United Kingdom has a double high water caused by the interaction between the region's different tidal harmonics, caused primarily by the east/west orientation of the English Channel and the fact that when it is high water at Dover it is low water at Land's End (some 300 nautical miles distant) and vice versa. This is contrary to the popular belief that the flow of water around the Isle of Wight creates two high waters. The Isle of Wight is important, however, since it is responsible for the 'Young Flood Stand', which describes the pause of the incoming tide about three hours after low water. Because the oscillation modes of the Mediterranean Sea and the Baltic Sea do not coincide with any significant astronomical forcing period, the largest tides are close to their narrow connections with the Atlantic Ocean. Extremely small tides also occur for the same reason in the Gulf of Mexico and Sea of Japan. Elsewhere, as along the southern coast of Australia, low tides can be due to the presence of a nearby amphidrome.

ANALYSIS

Isaac Newton's theory of gravitation first enabled an explanation of why there were generally two tides a day, not one, and offered hope for detailed understanding. Although it may seem that tides could be predicted via a sufficiently detailed knowledge of the instantaneous astronomical forcings, the actual tide at a given location is determined by astronomical forces accumulated over many days. Precise results require detailed knowledge of the shape of all the ocean basins-their bathymetry and coastline shape. Current procedure for analysing tides follows the method of harmonic analysis introduced in the 1860s by William Thomson. It is based on the principle that the astronomical theories of the motions of sun and moon determine a large number of component frequencies, and at each frequency there is a component of force tending to produce tidal motion, but that at each place of interest on the Earth, the tides respond at each frequency with an amplitude and phase peculiar to that locality.

At each place of interest, the tide heights are therefore measured for a period of time sufficiently long (usually more than a year in the case of a new port not previously studied) to enable the response at each significant tide-generating frequency to be distinguished by analysis, and to extract the tidal constants for a sufficient number of the strongest known components of the astronomical tidal forces to enable practical tide prediction. The tide heights are expected to follow the tidal force, with a constant amplitude and phase delay for each component. Because astronomical frequencies and phases can be calculated with certainty, the tide height at other times can then be predicted once the response to the harmonic components of the astronomical tide-generating forces has been found.

The main patterns in the tides are:

- the twice-daily variation
- the difference between the first and second tide of a day
- the spring-neap cycle
- the annual variation

The Highest Astronomical Tide is the perigean spring tide when both the sun and the moon are closest to the Earth. When confronted by a periodically varying function, the standard approach is to employ Fourier series, a form of analysis that uses sinusoidal functions as a basis set, having frequencies that are zero, one, two, three, etc. times the frequency of a particular fundamental cycle. These multiples are called harmonics of the fundamental frequency, and the process is termed harmonic analysis. If the basis set of sinusoidal functions suit the behaviour being modelled, relatively few harmonic terms need to be added. Orbital paths are very nearly circular, so sinusoidal variations are suitable for tides. For the analysis of tide heights, the Fourier series approach has in practice to be made more elaborate than the use of a single frequency and its harmonics. The tidal patterns are decomposed into many sinusoids having many fundamental frequencies, corresponding (as in the lunar theory) to many different combinations of the motions of the Earth, the moon, and the angles that define the shape and location of their orbits.

For tides, then, harmonic analysis is not limited to harmonics of a single frequency. In other words, the harmonies are multiples of many fundamental frequencies, not just of the fundamental frequency of the simpler Fourier series approach. Their representation as a Fourier series having only one fundamental frequency and its (integer) multiples would require many terms, and would be severely limited in the time-range for which it would be valid.

The study of tide height by harmonic analysis was begun by Laplace, William Thomson (Lord Kelvin), and George Darwin. A.T. Doodson extended their work, introducing the Doodson Number notation to organise the hundreds of resulting terms. This approach has been the international standard ever since, and the complications arise as follows: the tide-raising force is notionally given by sums of several terms. Each term is of the form

$$A \cdot \cos(w \cdot t + p)$$

where A is the amplitude, w is the angular frequency usually given in degrees per hour corresponding to t measured in hours, and p is the phase offset with regard to the astronomical state at time $t = 0$. There is one term for the moon and a second term for the sun. The phase p of the first harmonic for the moon term is called the lunitidal interval or high water interval. The next step is to accommodate the harmonic terms due to the elliptical shape of the orbits. Accordingly, the value of A is not a constant but also varying with time, slightly, about some average figure. Replace it then by A(t) where A is another sinusoid, similar to the cycles and epicycles of Ptolemaic theory. Accordingly,

$$A(t) = A \cdot (1 + Aa \cdot \cos(wa \cdot t + pa)) ,$$

which is to say an average value A with a sinusoidal variation about it of magnitude Aa , with frequency wa and phase pa . Thus the simple term is now the product of two cosine factors:

$$A \cdot [1 + Aa \cdot \cos(wa \cdot t + pa)] \cdot \cos(w \cdot t + p)$$

Given that for any x and y

$$\cos(x) \cdot \cos(y) = \tfrac{1}{2} \cdot \cos(x + y) + \tfrac{1}{2} \cdot \cos(x - y) ,$$

it is clear that a compound term involving the product of two cosine terms each with their own frequency is the same as three simple cosine terms that are to be added at the original frequency and also at frequencies which are the sum and difference of the two frequencies of the product term. (Three, not two terms, since the whole expression is $(1 + \cos(x)) \cdot \cos(y)$.) Consider further that the tidal force on a location depends also on whether the moon (or the sun) is above or below the plane of the equator, and that these attributes have their own periods also incommensurable with a day and a month, and it is clear that many combinations result. With a careful choice of the basic astronomical frequencies, the Doodson Number annotates the particular additions and differences to form the frequency of each simple cosine term.

Remember that astronomical tides do not include weather effects. Also, changes to local conditions (sandbank movement, dredging harbour mouths, etc.) away from those prevailing at the measurement time affect the tide's actual timing and magnitude. Organisations quoting a "highest astronomical tide" for some location may exaggerate the figure as a safety factor against analytical uncertainties, distance from the nearest measurement point, changes since the last observation time, ground subsidence, etc., to avert liability should an engineering work be overtopped. Special care is needed when assessing the size of a "weather surge" by subtracting the astronomical tide from the observed tide. Careful Fourier data analysis over a nineteen-year period (the National Tidal Datum Epoch in the U.S.) uses frequencies called the tidal harmonic constituents.

Nineteen years is preferred because the Earth, moon and sun's relative positions repeat almost exactly in the Metonic cycle of 19 years, which is long enough to include the 18.613-year lunar nodal tidal constituent. This analysis can be done using only the knowledge of the forcing period, but without detailed understanding of the mathematical derivation, which means that useful tidal tables have been constructed for centuries. The resulting amplitudes and phases can then be used to predict the expected tides. These are usually dominated by the constituents near 12 hours (the semi-diurnal constituents), but there are major constituents near 24 hours (diurnal) as well. Longer term constituents are 14 day or fortnightly, monthly, and semiannual. Semi-diurnal tides dominated coastline, but some areas such as the South China Sea and the Gulf of Mexico are primarily diurnal. In the semi-diurnal areas, the primary constituents M2 (lunar) and S2 (solar) periods differ slightly, so that the relative phases, and thus the amplitude of the combined tide, change fortnightly (14 day period).

In the M2 plot above, each cotidal line differs by one hour from its neighbors, and the thicker lines show tides in phase with equilibrium at Greenwich. The lines rotate around the amphidromic points counterclockwise in the northern hemisphere so that from Baja California Peninsula to Alaska and from France to Ireland the M2 tide propagates northward. In the southern hemisphere this direction is clockwise. On the other hand M2 tide propagates counterclockwise around New Zealand, but this is because the islands act as a dam and permit the tides to have different heights on the islands' opposite sides. (The tides do propagate northward on the east side and southward on the west coast, as predicted by theory.)

The exception is at Cook Strait where the tidal currents periodically link high to low water. This is because cotidal lines 180° around the amphidromes are in opposite phase, for example high water across from low water at each end of Cook Strait. Each tidal constituent has a different pattern of amplitudes, phases, and amphidromic points, so the M2 patterns cannot be used for other tide components.

EXAMPLE CALCULATION

Because the moon is moving in its orbit around the earth and in the same sense as the Earth's rotation, a point on the earth must rotate slightly further to catch up so that the time between semidiurnal tides is not twelve but 12.4206 hours-a bit over twenty-five minutes extra. The two peaks are not equal. The two high tides a day alternate in maximum heights: lower high (just under three feet), higher high (just over three feet), and again lower high. Likewise for the low tides. When the Earth, moon, and sun are in line (sun-Earth-moon, or sun-moon-Earth) the two main influences combine to produce spring tides; when the two forces are opposing each other as when the angle moon-Earth-sun is close to ninety degrees, neap tides result. As the moon moves around its orbit it changes from north of the equator to south of the equator. The alternation in high tide heights becomes smaller, until they are the same (at the lunar equinox, the moon is above the equator), then redevelop but with the other polarity, waxing to a maximum difference and then waning again.

CURRENT

The tides' influence on current flow is much more difficult to analyse, and data is much more difficult to collect. A tidal height is a simple number which applies to a wide region simultaneously. A flow has both a magnitude and a direction, both of which can vary substantially with depth and over short distances due to local bathymetry. Also, although a water channel's center is the most useful measuring site, mariners object when current-measuring equipment obstructs waterways. A flow proceeding up a curved channel is the same flow, even though its direction varies continuously along the channel. Surprisingly, flood and ebb flows are often not in opposite directions. Flow direction is determined by the upstream channel's shape, not the downstream channel's shape. Likewise, eddies may form in only one flow direction.

Nevertheless, current analysis is similar to tidal analysis: in the simple case, at a given location the flood flow is in mostly one direction, and the ebb flow in another direction. Flood velocities are given positive sign, and ebb velocities negative sign. Analysis proceeds as though these are tide heights. In more complex situations, the main ebb and flood flows do not dominate. Instead, the flow direction and magnitude trace an ellipse over a tidal cycle (on a polar plot) instead of along the ebb and flood lines. In this case, analysis might proceed along pairs of directions, with the primary and secondary directions at right angles. An alternative is to treat the tidal flows as complex numbers, as each value has both a magnitude and a direction.

Tide flow information is most commonly seen on nautical charts, presented as a table of flow speeds and bearings at hourly intervals, with separate tables for spring and neap tides. The timing is relative to high water at some harbour where the tidal behaviour is similar in pattern, though it may be far away. As with tide height predictions, tide flow predictions based only on astronomical factors do not incorporate weather conditions, which can completely change the outcome. The tidal flow through Cook Strait between the two main islands of New Zealand is particularly interesting, as the tides on each side of the strait are almost exactly out of phase, so that one side's high water is simultaneous with the other's low water. Strong currents result, with almost zero tidal height change in the strait's center. Yet, although the tidal surge normally flows in one direction for six hours and in the reverse direction for six hours, a particular surge might last eight or ten hours with the reverse surge enfeebled. In especially boisterous weather conditions, the reverse surge might be entirely overcome so that the flow continues in the same direction through three or more surge periods.

A further complication for Cook Strait's flow pattern is that the tide at the north side (e.g. at Nelson) follows the common bi-weekly spring-neap tide cycle (as found along the west side of the country), but the south side's tidal pattern has only one cycle per month, as on the east side: Wellington, and Napier. The graph of Cook Strait's tides shows separately the high water and low water height and time, through November 2007; these are not measured values but instead are calculated from tidal parameters derived from years-old measurements. Cook Strait's nautical chart offers tidal current information. For instance the January 1979 edition for 41°13·9'S 174°29·6'E (north west of Cape Terawhiti) refers timings to Westport while the January 2004 issue refers to Wellington.

Near Cape Terawhiti in the middle of Cook Strait the tidal height variation is almost nil while the tidal current reaches its maximum, especially near the notorious Karori Rip. Aside from weather effects, the actual currents through Cook Strait are influenced by the tidal height differences between the two ends of the strait and as can be seen, only one of the two spring tides at the north end (Nelson) has a counterpart spring tide at the south end (Wellington), so the resulting behaviour follows neither reference harbour.

POWER GENERATION

Tidal energy can be extracted by two means: inserting a water turbine into a tidal current, or building ponds that release/admit water through a turbine. In the first case, the energy amount is entirely determined by the timing and tidal current magnitude. However, the best currents may be unavailable because the turbines would obstruct ships. In the second, the impoundment dams are expensive to construct, natural water cycles are completely disrupted, ship navigation is disrupted. However, with multiple ponds, power can be generated at chosen times.

So far, there are few installed systems for tidal power generation (most famously, La Rance at Saint Malo, France) which face many difficulties. Aside from environmental issues, simply withstanding corrosion and biological fouling pose engineering challenges. Tidal power proponents point out that, unlike wind power systems, generation levels can be reliably predicted, save for weather effects. While some generation is possible for most of the tidal cycle, in practice turbines lose efficiency at lower operating rates. Since the power available from a flow is proportional to the cube of the flow speed, the times during which high power generation is possible are brief.

NAVIGATION

Tidal flows are important for navigation, and significant errors in position occur if they are not accommodated. Tidal heights are also important; for example many rivers and harbours have a shallow "bar" at the entrance which prevents boats with significant draft from entering at low tide. Until the advent of automated navigation, competence in calculating tidal effects was important to naval officers. The certificate of examination for lieutenants in the Royal Navy once declared that the prospective officer was able to "shift his tides".

Tidal flow timings and velocities appear in tide charts or a tidal stream atlas. Tide charts come in sets. Each chart covers a single hour between one high water and another (they ignore the leftover 24 minutes) and show the average tidal flow for that hour. An arrow on the tidal chart indicates the direction and the average flow speed (usually in knots) for spring and neap tides. If a tide chart is not available, most nautical charts have "tidal diamonds" which relate specific points on the chart to a table giving tidal flow direction and speed.

The standard procedure to counteract tidal effects on navigation is to (1) calculate a "dead reckoning" position (or DR) from travel distance and direction, (2) mark the chart (with a vertical cross like a plus sign) and (3) draw a line from the DR in the tide's direction. The distance the tide moves the boat along this line is computed by the tidal speed, and this gives an "estimated position" or EP (traditionally marked with a dot in a triangle). Nautical charts display the water's "charted depth" at specific locations with "soundings" and the use of bathymetric contour lines to depict the submerged surface's shape. These depths are relative to a "chart datum", which is typically the water level at the lowest possible astronomical tide (although other datums are commonly used, especially historically, and tides may be lower or higher for meteorological reasons) and are therefore the minimum possible water depth during the tidal cycle. "Drying heights" may also be shown on the chart, which are the heights of the exposed seabed at the lowest astronomical tide.

Tide tables list each day's high and low water heights and times. To calculate the actual water depth, add the charted depth to the published tide height. Depth for other times can be derived from tidal curves published for major ports. The rule of twelfths can suffice if an accurate curve is not available. This approximation presumes that the increase in depth in the six hours between low and high water is: first hour - 1/12, second - 2/12, third - 3/12, fourth - 3/12, fifth - 2/12, sixth - 1/12.

BIOLOGICAL ASPECTS

INTERTIDAL ECOLOGY

Intertidal ecology is the study of intertidal ecosystems, where organisms live between the low and high water lines. At low water, the intertidal is exposed (or 'emersed') whereas at high water, the intertidal is underwater (or 'immersed'). Intertidal ecologists therefore study the interactions between intertidal organisms and their environment, as well as among the different species. The most important interactions may vary according to the type of intertidal community. The broadest classifications are based on substrates - rocky shore or soft bottom.

Intertidal organisms experience a highly variable and often hostile environment, and have adapted to cope with and even exploit these conditions. One easily visible feature is vertical zonation, in which the community divides into distinct horizontal bands of specific species at each elevation above low water. A species' ability to cope with desiccation determines its upper limit, while competition with other species sets its lower limit.

Humans use intertidal regions for food and recreation. Overexploitation can damage intertidals directly. Other anthropogenic actions such as introducing invasive species and climate change have large negative effects. Marine Protected Areas are one option communities can apply to protect these areas and aid scientific research.

BIOLOGICAL RHYTHMS

The approximately fortnightly tidal cycle has large effects on intertidal and marine organisms. Hence their biological rhythms tend to occur in rough multiples of this period. Many other animals such as the vertebrates, display similar rhythms. Examples include gestation and egg hatching. In humans, the menstrual cycle lasts roughly a lunar month, an even multiple of the tidal period. Such parallels at least hint at the common descent of all animals from a marine ancestor.

OTHER TIDES

When oscillating tidal currents in the stratified ocean flow over uneven bottom topography, they generate internal waves with tidal frequencies. Such waves are called internal tides.Shallow areas in otherwise open water can experience rotary tidal currents, flowing in directions that continually change and thus the flow direction (not the flow) completes a full rotation in 12½ hours (for example, the Nantucket Shoals).

In addition to oceanic tides, large lakes can experience small tides and even planets can experience atmospheric tides and Earth tides. These are continuum mechanical phenomena. The first two take place in fluids. The third affects the Earth's thin solid crust surrounding its semi-liquid interior (with various modifications).

LAKE TIDES

Large lakes such as Superior and Erie can experience tides of 1 to 4 cm, but these can be masked by meteorologically induced phenomena such as seiche. The tide in Lake Michigan is described as 0.5 to 1.5 inches (13 to 38 mm) or 1¾ inches.

ATMOSPHERIC TIDES

Atmospheric tides are negligible at ground level and aviation altitudes, masked by weather's much more important effects. Atmospheric tides are both gravitational and thermal in origin and are the dominant dynamics from about 80 to 120 kilometres (50 to 75 mi), above which the molecular density becomes too low to support fluid behavior.

EARTH TIDES

Earth tides or terrestrial tides affect the entire Earth's mass, which acts similarly to a liquid gyroscope with a very thin crust. The Earth's crust shifts (in/out, east/west, north/south) in response to lunar and solar gravitation, ocean tides, and atmospheric loading. While negligible for most human activities, terrestrial tides' semi-diurnal amplitude can reach about 55 centimetres (22 in) at the equator-15 centimetres (5.9 in) due to the sun-which is important in GPS calibration and VLBI measurements. Precise astronomical angular measurements require knowledge of the Earth's rotation rate and nutation, both of which are influenced by Earth tides. The semi-diurnal M2 Earth tides are nearly in phase with the moon with a lag of about two hours. Some particle physics experiments must adjust for terrestrial tides. For instance, at CERN and SLAC, the very large particle accelerators account for terrestrial tides. Among the relevant effects are circumference deformation for circular accelerators and particle beam energy. Since tidal forces generate currents in conducting fluids in the Earth's interior, they in turn affect the Earth's magnetic field. Earth tides have also been linked to the triggering of earthquakes.

GALACTIC TIDES

Galactic tides are the tidal forces exerted by galaxies on stars within them and satellite galaxies orbiting them. The galactic tide's effects on the Solar System's Oort cloud are believed to cause 90 percent of long-period comets.

MISAPPLICATIONS

Tsunamis, the large waves that occur after earthquakes, are sometimes called tidal waves, but this name is given by their resemblance to the tide, rather than any actual link to the tide. Other phenomena unrelated to tides but using the word tide are rip tide, storm tide, hurricane tide, and black or red tides.

TIDAL RESONANCE

In oceanography, a tidal resonance occurs when the tide excites one of the resonant modes of the ocean . The effect is most striking when a continental shelf is about a quarter wavelength wide. Then an incident tidal wave can be reinforced by reflections between the coast and the shelf edge, the result producing a much higher tidal range at the coast.

Famous examples of this effect are found in the Bay of Fundy, where the world's highest tides are reportedly found, and in the Bristol Channel. Large tides due to resonances are also found on the Patagonian Shelf and on the N.W. Australian continental shelf. The speed of long waves in the ocean is given, to a good approximation, by $\sqrt{gh}$, where g is the acceleration of gravity and h is the depth of the ocean . For a typical continental shelf with a depth of 100 m, the speed is approximately 30 m/s. So if the tidal period is 12 hours, a quarter wavelength shelf will have a width of about 300 km.

With a narrower shelf, there is still a resonance but it has less effect at tidal frequencies. However the effect is still enough to partly explain why tides along a coast lying behind a continental shelf are often higher than at offshore islands in the deep ocean. The strong tidal currents associated with resonances also mean that the resonant regions are the areas where most tidal energy is dissipated.

In the deep ocean, where the depth is typically 4000 m, the speed of long waves increases to approximately 200 m/s. The difference in speed, when compared to the shelf, is responsible for the reflections at the continental shelf edge. Away from resonance this can stop tidal energy moving onto the shelf. However near a resonant frequency the phase relationships between the wave on the shelf and in the deep ocean can have the effect of drawing energy onto the shelf.

The increased speed of long waves in the deep ocean means that the tidal wavelength there is of order 10,000 km. As the ocean basins have a similar size, they also have the potential of being resonant . In practice deep ocean resonances are difficult to observe, probably because the deep ocean loses tidal energy too rapidly to the resonant shelves.

TIDAL LOCKING

The above concept of tidal resonance differs from another sort of resonance resulting from tides, called tidal locking, which causes a moon's rotational period to coincide with the period of its revolution around the planet that it orbits, so that one side of the moon always faces the planet.

TIDAL RANGE

The tidal range is the vertical difference between the high tide and the succeeding low tide. Tides are the rise and fall of sea levels caused by the combined effects of the gravitational forces exerted by the Moon and the Sun and the rotation of the Earth. The tidal range is not constant, but changes depending on where the sun and the moon are.

The most extreme tidal range occurs around the time of the full or new moons, when the gravitational forces of both the Sun and Moon are in phase, reinforcing each other in the same direction (new moon), or are exactly the opposite phase (full). This type of tide is known as a spring tide. During neap tides, when the Moon and Sun's gravitational force vectors act in quadrature (making a right angle to the Earth's orbit), the difference between high and low tides is smaller. Neap tides occur during the first and last quarters of the moon's phases. The largest annual tidal range can be expected around the time of the equinox, if coincidental with a spring tide. Tidal data for coastal areas is published by the national hydrographic service of the country concerned. Tidal data is based on astronomical phenomena and is predictable. Storm force winds blowing from a steady direction for a prolonged time interval combined with low barometric pressure can increase the tidal range, particularly in narrow bays. Such weather-related effects on the tide, which can cause ranges in excess of predicted values and can cause localized flooding, are not calculable in advance.

GEOGRAPHY

The typical tidal range in the open ocean is about 0.6 metres (2 feet). Closer to the coast, this range is much greater. Coastal tidal ranges vary globally and can differ anywhere from near zero to over 11 metres (38 feet). The exact range depends on the volume of water adjacent to the coast, and the geography of the basin the water sits in. Larger bodies of water have higher ranges, and the geography can act as a funnel amplifying or dispersing the tide. The world's largest tidal range of 11.7 metres (38.4 feet) occurs at Burntcoat Head in the Bay of Fundy, Eastern Canada. The Bristol Channel, between England and Wales in the United Kingdom, regularly experiences tidal ranges of up to 9 metres. The top 50 locations with the largest tidal ranges world-wide are listed by the National Oceanic and Atmospheric Administration of the United States. Some of the smallest tidal ranges occur in the Mediterranean, Baltic, and Caribbean Seas. A point within a tidal system where the tidal range is almost zero is called an amphidromic point.

CLASSIFICATION

The tidal range has been classified as:

- Micromareal, when the tidal range is lower than 2 metres.
- Mesomareal, when the tidal range is between 2 metres and 4 metres.
- Macromareal, when the tidal range is higher than 4 metres.

TIDAL PRISM

A tidal prism is the volume of water in an estuary or inlet between mean high tide and mean low tide. or the volume of water leaving an estuary at ebb tide.

The inter-tidal prism volume can be expressed by the relationship: P=H A, where H is the average tidal range and A is the average surface area of the basin. It can also be thought of as the volume of the incoming tide plus the river discharge. Simple tidal prism models stated the relationship of river discharge and inflowing ocean water as Prism=Volume of ocean water coming into an estuary on the flood tide + Volume of river discharge mixing with that ocean water; however, there is some controversy as to whether traditional prism models are accurate. The size of an estuary's tidal prism is dependent on the basin of that estuary, the tidal range and other frictional forces.

APPLICATIONS OF TIDAL PRISM

Calculations of tidal prism are useful in determining the residence time of water (and pollutants) in an estuary. If it is known how much water is exported compared to how much of the estuarine water remains, it can be determined how long pollutants reside in that estuary. If the tidal prism forms a large proportion of the water in an estuary at high tide, then when the tide ebbs, it will take with it the majority of the water (this occurs in shallow estuaries) and any pollutants or sediments suspended in that water.

This means that the estuary has a good flushing time, or that the residence time of water in that estuary is low. On the contrary, in deeper estuaries, the amount of water that is influenced by the tides forms a smaller proportion of the total water. The difference between high tide and low tide is not as great as in shallower estuaries creating a smaller tidal prism and a longer residence time. The size of an inlet or estuary is determined, according to O'Brien by tidal prism. Tidal prism magnitude can be calculated by multiplying the area of the estuary by the tidal range of that estuary.

During spring or fall tides, when sea level is relatively high and floods backbarrier areas that are normally above tidal inundation, the cross sectional area at the entrance of the estuary increases as tidal prism increases. Since tidal prism is largely a function of area of open water and tidal range, it can be changed by alterations of the basin area of estuaries and inlets as in dredging; however, if the estuary or inlet is dredged, or the size changed, the channel will fill in with sediment until it has returned to tidal prism equilibrium.

SAND TRANSPORT

Additionally, there are correlations between tidal prism and amount of sediment deposited and exported in an estuary or inlet. The Walton and Adams relationship shows a strong relationship between the magnitude of the tidal prism and the volume of sand in ebb dominated deltas. The larger the tidal prism, the larger the amount of sand that is deposited in deltas in ebb-dominated estuaries. Inlets with small tidal prisms have too little power to remove sand deposited from adjacent shores. Inlets with large tidal prisms can erode sand and deposit it in ebb-tidal deltas in deeper waters (National Research Council). The size of ebb tidal deltas is proportional to tidal prism. If tidal prism increases, there is an increase in deltas and shoals formed by sand transport during ebb tide.

TIDAL PRISM MODELS AND ASSUMPTIONS

There are assumptions that go along with tidal prism models. The first is that they are applied to smaller estuaries (less than a few kilometers wide) and secondly, that the estuaries are internally well mixed. Additionally, it is assumed that the water entering the estuary is of oceanic salinity mixing with the fresh river discharge, and that the mixed water will be exported on the ebb tide.

Officer provides a model for simple tidal prism theory where the estuary is represented by a box with the inflow as the volume of river discharge at a salinity of 0, within the estuary, the river discharge mixes with the volume of the tide flooding in (Vp) from the ocean at oceanic salinity (So) and the mixed VR + VP) water flows out at ebb tide.

TIDE-PREDICTING MACHINE

A tide-predicting machine was a special-purpose mechanical analog computer of the late 19th and early 20th centuries, constructed and set up to predict the ebb and flow of sea tides and the irregular variations in their heights - which change in mixtures of rhythms, that never (in the aggregate) repeat themselves exactly. Its purpose was to shorten the laborious and error-prone computations of tide-prediction. Such machines usually provided predictions valid from hour to hour and day to day for a year or more ahead. The first tide-predicting machine, designed and built in 1872-3, and followed by two larger machines on similar principles in 1876 and 1879, was conceived by Sir William Thomson (who later became Lord Kelvin). Thomson had previously, during the 1860s, introduced the method of harmonic analysis of tidal patterns. The first machine was designed by Thomson with the collaboration of Edward Roberts (assistant at the UK HM Nautical Almanac Office), and of Alexander Légé, who constructed it.

In the US, another tide-predicting machine on a different pattern (shown right) was designed by William Ferrel and built in 1881-2. Developments and improvements continued in the UK, US and Germany through the first half of the 20th century. The machines became widely used for constructing official tidal predictions for general marine navigation. They came to be regarded as of military strategic importance during World War I, and again during the second World War, when the US No.2 Tide Predicting Machine, described below, was classified, along with the data that it produced, and used to predict tides for the D-day Normandy landings and all the island landings in the Pacific war. Military interest in such machines continued even for some time afterwards. They were made obsolete by digital electronic computers that can be programmed to carry out similar computations, but the tide-predicting machines continued in use until the 1960s and 1970s. Several examples of tide-predicting machines remain on display as museum pieces, occasionally put into operation for demonstration purposes, monuments to the mathematical and mechanical ingenuity of their creators.

BACKGROUND TO THE PROBLEM SOLVED BY THE MACHINES

Modern scientific study of tides dates back to Isaac Newton's 'Principia' of 1687, in which he applied the theory of gravitation to make a first approximation of the effects of the Moon and Sun on the Earth's tidal waters. The approximation developed by Newton and his successors of the next 90 years is known as the 'equilibrium theory' of tides. Beginning in the 1770s, Pierre-Simon Laplace made a fundamental advance on the equilibrium approximation by bringing into consideration non-equilibrium dynamical aspects of the motion of tidal waters that occurs in response to the tide-generating forces due to the Moon and Sun.

Laplace's improvements in theory were substantial, but they still left prediction in an approximate state. This position changed in the 1860s when the local circumstances of tidal phenomena were more fully brought into account by William Thomson's application of Fourier analysis to the tidal motions. Thomson's work in this field was then further developed and extended by George Darwin: Darwin's work was based on the lunar theory current in his time. His symbols for the tidal harmonic constituents are still used. Darwin's harmonic developments of the tide-generating forces were later brought by A T Doodson up to date and extended in light of the new and more accurate lunar theory of E W Brown that remained current through most of the twentieth century.

The state to which the science of tide-prediction had arrived by the 1870s can be summarized: Astronomical theories of the Moon and Sun had identified the frequencies and strengths of different components of the tide-generating force. But effective prediction at any given place called for measurement of an adequate sample of local tidal observations, to show the local tidal response at those different frequencies, in amplitude and phase. Those observations had then to be analyzed, to derive the coefficients and phase angles.

Then, for purposes of prediction, those local tidal constants had to be recombined, each with a different component of the tide-generating forces to which it applied, and at each of a sequence of future dates and times, and then the different elements finally collected together to obtain their aggregate effects. In the age when calculations were done by hand and brain, with pencil and paper and tables, this was recognized as an immensely laborious and error-prone undertaking. Thomson recognized that what was needed was a convenient and preferably automated way to evaluate repeatedly the sum of tidal terms such as:

$$A_1 \cos(z_1 t \;.\; i_1) \;.\; A_2 \cos(z_2 t \;.\; i_2) \;.\; A_3 \cos(z_3 t \;.\; i_3) \;.\; \hat{}$$

containing 10, 20 or even more trigonometrical terms, so that the computation could conveniently be repeated in full for each of a very large number of different chosen values of the date/time t. This was the core of the problem solved by the tide-predicting machines.

HOW THEY WORKED TO PREDICT THE TIDES

Thomson conceived his aim as to construct a mechanism that would evaluate this trigonometrical sum physically, e.g. as the vertical position of a pen that could then plot a curve on a moving band of paper. There were several mechanisms available to him for converting rotary motion into sinusoidal motion. One of them is shown in the schematic (right). A rotating drive-wheel is fitted with an off-center peg. A shaft with a horizontally-slotted section is free to move vertically up and down. The wheel's off-center peg is located in the slot.

As a result, when the peg moves around with the wheel, it can make the shaft move up and down within limits. This arrangement shows that when the drive-wheel rotates uniformly, say clockwise, the shaft moves sinusoidally up and down. The vertical position of the center of the slot, at any time t, can then be expressed as $A_1 \cos(z_1 t \;.\; i_1)$, where A is the radial distance from the wheel's center to the peg, z_1 is the rate at which the wheel turns (in radians per unit of time), and i_1 is the starting phase angle of the peg, measured in radians from the 12 o'clock position to the angular position where the peg was at time zero. This arrangement makes a physical analog of just one trigonometrical term. Thomson needed to construct a physical sum of many such terms.

At first he inclined to use gears. Then he discussed the problem with engineer Beauchamp Tower before the British Association meeting in 1872, and Tower suggested the use of a device that (as he remembered) was once used by Wheatstone. It was a chain running alternately over and under a sequence of pulleys on movable shafts. The chain was fixed at one end, and the other (free) end was weighted to keep it taut. As each shaft moved up or down it would take up or release a corresponding length of the chain. The movements in position of the free (movable) end of the chain represented the sum of the movements of the different shafts. The movable end was kept taut, and fitted with a pen and a moving band of paper on which the pen plotted a tidal curve. In some designs, the movable end of the line was connected instead to a dial and scale from which tidal heights could be read off. A long cord, with one end held fixed, passed vertically upwards and over a first upper pulley, then vertically downwards and under the next, and so on. These pulleys were all moved up and down by cranks, and each pulley took in or let out cord according to the direction in which it moved. These cranks were all moved by trains of wheels gearing into the wheels fixed on a drive shaft. The greatest number of teeth on any wheel was 802 engaging with another of 423. All the other wheels had comparatively small numbers of teeth. A flywheel of great inertia enabled the operator to turn the machine fast, without jerking the pulleys, and so to run off a year's curve in about twenty-five minutes.

Thomson acknowledged that the use of an over-and-under arrangement of the flexible line that summed the motion components was suggested to him in August 1872 by engineer Beauchamp Tower.

ONLINE DEMONSTRATION OF THE MECHANISM

An online demonstration is available to show the principle of operation of a 7-component version of a tide-predicting machine otherwise like Thomson's (Kelvin's) original design.

The animation shows part of the operation of the machine: the motions of several pulleys can be seen, each moving up and down to simulate one of the tidal frequencies; and the animation also shows how these sinusoidal motions were generated by wheel rotations and how they were combined to form the resulting tidal curve.

Not shown in the animation is the way in which the individual motions were generated in the machine at the correct relative frequencies, by gearing in the correct ratios, or how the amplitudes and starting phase angles for each motion were set in an adjustable way.

These amplitudes and starting phase angles represented the local tidal constants, separately reset, and different for each place for which predictions were to be made. Also, in the real Thomson machines, to save on motion and wear of the other parts, the shaft and pulley with the largest expected motion (for the M2 tide component at twice per lunar day) was mounted nearest to the pen, and the shaft and pulley representing the smallest component was at the other end, nearest to the point of fixing of the flexible cord or chain, to minimize unnecessary motion in the most part of the flexible cord.

HISTORY OF THEIR BUILDING AND USE

The first tide-predicting machines 1872-1883

The first tide predicting machine, designed in 1872 and of which a model was exhibited at the British Association meeting in 1873 (for computing 8 tidal components), followed in 1875-6 by a machine on a slightly larger scale (for computing 10 tidal components), was designed by Sir William Thomson (who later became Lord Kelvin). The 10-component machine and results obtained from it were shown at the Paris Exhibition in 1878. An enlarged and improved version of the machine, for computing 20 tidal components, was built for the Government of India in 1879, and then modified in 1881 to extend it to compute 24 harmonic components. In these machines, the prediction was delivered in the form of a continuous graphical pen-plot of tidal height against time. The plot was marked with hour- and noon-marks, and was made by the machine on a moving band of paper as the mechanism was turned. A year's tidal predictions for a given place, usually a chosen seaport, could be plotted by the 1876 and 1879 machines in about four hours (but the drives had to be rewound during that time).

In 1881-2, another tide predicting machine, operating quite differently, was designed by William Ferrel and built in Washington under Ferrel's direction by E G Fischer (who later designed the successor machine described below, which was in operation at the US Coast and Geodetic Survey from 1912 until the 1960s). Ferrel's machine delivered predictions by telling the times and heights of successive high and low waters, shown by pointer-readings on dials and scales. These were read by an operator who copied the readings on to forms, to be sent to the printer of the US tide-tables.

These machines had to be set with local tidal constants special to the place for which predictions were to be made. Such numbers express the local tidal response to individual components of the global tide-generating potential, at different frequencies. This local response, shown in the timing and the height of tidal contributions at different frequencies, is a result of local and regional features of the coasts and sea-bed. The tidal constants are usually evaluated from local histories of tide-gauge observations, by harmonic analysis based on the principal tide-generating frequencies as shown by the global theory of tides and the underlying lunar theory.

Thomson was also responsible for originating the method of harmonic tidal analysis, and for devising a harmonic analyzer machine, which partly mechanized the evaluation of the constants from the gauge readings. Development and improvement based on the experience of these early machines continued through the first half of the 20th century.

British Tide Predictor No.2, after initial use to generate data for Indian ports, was used for tide prediction for the British empire beyond India, and transferred to the National Physical Laboratory in 1903. British Tide Predictor No.3 was sold to the French Government in 1900 and used to generate French tide tables. US Tide Predicting Machine No.2 ("Old Brass Brains") was designed in the 1890s, completed and brought into service in 1912, used for several decades including during the second World War, and retired in the 1960s. Tide-predicting machines were built in Germany during World War I, and again in the period 1935-8.

Two of the last to be built were:

- A TPM built in 1947 for the Norwegian Hydrographic Service by Chadburn of Liverpool, and designed to compute 30 tidal harmonic

constituents; used until 1975 to compute official Norwegian Tide Tables, before being superseded by digital computing.
- The Doodson-Légé TPM built in 1949,
- An East German TPM built 1953-5.

Tide predicting machines on display

They can be seen in London, Washington, Liverpool, and elsewhere, including the Deutsches Museum in Munich.

12

Ocean Environment Engineering

COASTAL MANAGEMENT

In some jurisdictions the terms sea defense and coastal protection are used to mean, respectively, defense against flooding and erosion. The term coastal defence is the more traditional term, but coastal management has become more popular as the field has expanded to include techniques that allow erosion to claim land.

HISTORICAL BACKGROUND

Coastal engineering, as it relates to harbours, starts with the development of ancient civilizations together with the origin of maritime traffic, perhaps before 3500 B.C. Docks, breakwaters, and other harbour works were built by hand and often in a grand scale. Basic source of modern literature on coastal engineering is the "European Code of Conduct for Coastal Zones" issued by the European Council in 1999. This document was prepared by the Group of Specialists on Coastal Protection and should be used 'as a source of inspiration for national legislation and practice' by decision makers. The Group of Specialists on Coastal Protection (PE-S-CO), set up in 1995, pursuant to a decision by the Committee of Ministers of the Council of Europe, met for the first time on 6 and 7 June 1996.

It noted that a great deal of technical and scientific research had been carried out in the field of coastal protection and that various principles and legal texts had been drawn up. It also noted that all of the work undertaken highlighted the need for integrated management and planning of coastal areas, but that, despite all the efforts already made, the situation of coastal areas continued to deteriorate.

The Group had acknowledged that this was due to difficulties in implementing the concept of "integrated management", and that it was becoming necessary to provide instruments which would make it easier to apply the principles of integrated coastal management and planning, which had to be pursued to ensure sustainable management of coastal areas. The Group therefore proposed that the Council of Europe, in close co-operation with the European Union for Coastal Conservation (EUCC) and United Nations Environment Programme (UNEP). The final version of the Code as well of the a MODEL LAW to be used as a guide for modefying local and national legislation, can be free downloaded from the web.

Ancient harbour works are still visible in a few of the harbours that exist today, while others have recently been explored by underwater archaeologists. Most of the grander ancient harbor works have disappeared following the fall of the Roman Empire. Most ancient coastal efforts were directed to port structures, with the exception of a few places where life depended on coastline protection. Venice and its lagoon is one such case. Protection of the shore in Italy, England and the Netherlands can be traced back at least to the 6th century. The ancients understood such phenomena as the Mediterranean currents and wind patterns and the wind-wave cause-effect link.

The Romans introduced many revolutionary innovations in harbor design. They learned to build walls underwater and managed to construct solid breakwaters to protect fully exposed harbors. In some cases wave reflection may have been used to prevent silting. They also used low, water-surface breakwaters to trip the waves before they reached the main breakwater. They became the first dredgers in the Netherlands to maintain the harbour at Velsen. Silting problems here were solved when the previously sealed solid piers were replaced with new "open"-piled jetties.

Middle Ages

The threat of attack from the sea caused many coastal towns and their harbours to be abandoned. Other harbours were lost due to natural causes such as rapid silting, shoreline advance or retreat, etc. The Venetian Lagoon was one of the few populated coastal areas with continuous prosperity and development where written reports document the evolution of coastal protection works. Engineering and scientific skills remained alive in the east, in Byzantium, where the Eastern Roman Empire survived for six hundred years while Western Rome decayed.

Modern Age

Although great strides were made in the general scientific arena, little improvement was done beyond the Roman approach to harbour construction after the Renaissance. In the early 19th century, the advent of the steam engine, the search for new lands and trade routes, the expansion of the British Empire through her colonies, and other influences, all contributed to the revitalization of sea trade and a renewed interest in port works.

Twentieth century

Evolution of shore protection and the shift from structures to beach nourishment. Prior to the 1950s, the general practice was to use hard structures to protect against beach erosion or storm damages. These structures were usually coastal armoring such as seawalls and revetments or sand-trapping structures such as groynes. During the 1920s and '30s, private or local community interests protected many areas of the shore using these techniques in a rather ad hoc manner. In certain resort areas, structures had proliferated to such an extent that the protection actually impeded the recreational use of the beaches. Erosion of the sand continued, but the fixed back-beach line remained, resulting in a loss of beach area.

The obtrusiveness and cost of these structures led in the late 1940s and early 1950s, to move toward a new, more dynamic, method. Projects no longer relied solely on hard coastal defence structures, as techniques were developed which replicated the protective characteristics of natural beach and dune systems. The resultant use of artificial beaches and stabilized dunes as an engineering approach was an economically viable and more environmentally friendly means for dissipating wave energy and protecting coastal developments.

Over the past hundred years the limited knowledge of coastal sediment transport processes at the local authorities level has often resulted in inappropriate measures of coastal erosion mitigation. In many cases, measures may have solved coastal erosion locally but have exacerbated coastal erosion problems at other locations -up to tens of kilometers away- or have generated other environmental problems.

CURRENT CHALLENGES IN COASTAL MANAGEMENT

The coastal zone is a dynamic equilibrium area of natural change and of increasing human use. They occupy less than 15% of the Earth's land surface; yet accommodate more than 40% of the world population (it is estimated that 3.1 billion people live within 200 kilometres from the sea). With three-quarters of the world population expected to reside in the coastal zone by 2025, human activities originating from this small land area will impose an inordinate amount of pressures on the global system. Coastal zones contain rich resources to produce goods and services and are home to most commercial and industrial activities. In the European Union, almost half of the population now lives within 50 kilometres of the sea and coastal zone resources produce much of the Union's economic wealth.

The fishing, shipping and tourism industries all compete for vital space along Europe's estimated 89 000 kilometres of coastline, and coastal zones contain some of Europe's most fragile and valuable natural habitats. Shore protection consists up to the 50's of interposing a static structure between the sea and the land to prevent erosion and or flooding, and it has a long history. From that period new technical or friendly policies have been developed to preserve the environment when possible. Is already important where there are extensive low-lying areas that require protection. For instance: Venice,

New Orleans, Nagara river in Japan, the Netherlands, Caspian Sea Protection against the sea level rise in the 21st century will be especially important, as sea level rise is currently accelerating. This will be a challenge to coastal management, since seawalls and breakwaters are generally expensive to construct, and the costs to build protection in the face of sea-level rise would be enormous.

Changes on sea level have a direct adaptative response from beaches and coastal systems, as we can see in the succession of a lowering sea level. When the sea level rises, coastal sediments are in part pushed up by wave and tide energy, so sea-level rise processes have a component of sediment transport landwards. This results in a dynamic model of rise effects with a continuous sediment displacement that is not compatible with static models where coastline change is only based on topographic data.

PLANNING APPROACHES

There are five generic strategies for coastal defense:

- inaction leading to eventual abandonment
- Managed retreat or realignment, which plans for retreat and adopts engineering solutions that recognise natural processes of adjustment, and identifies a new line of defence where to construct new defences
- Hold the line, shoreline protection, whereby seawalls are constructed around the coastlines
- Move seawards, this happens by constructing new defenses seaward the original ones
- Limited intervention, accommodation, by which adjustments are made to be able to cope with inundation, raising coastal land and buildings vertically

The decision to choose a strategy is site-specific, depending on pattern of relative sea-level change, geomorphological setting, sediment availability and erosion, as well a series of social, economic and political factors.

Alternatively, integrated coastal zone management approaches may be used to prevent development in erosion- or flood-prone areas to begin with. Growth management can be a challenge for coastal local authorities who often struggle to provide the infrastructure required by new residents seeking seachange lifestyles. Sustainable transport investment to reduce the average footprint of coastal visitors is often a good way out of coastal gridlock. Examples include Dongtan and the Gold Coast Oceanway.

The 'Managed Retreat' option, involving no protection, is cheap and expedient. The coast takes care of itself and coastal facilities are abandoned to coastal erosion, with either gradual landward retreat or evacuation and resettlement elsewhere. This is the usual response when land of little value will be lost. The only pollution produced is from the resettlement process. Where endangered property has high value, it is less often applied.

MANAGED RETREAT

Managed retreat is an alternative to constructing or maintaining coastal structures. Managed retreat allows an area to become flooded. This process is usually in low lying estuarine or deltaic areas and floods land that has at some point in the past been reclaimed from the sea. Managed retreat is often a response to a change in sediment budget or to sea level rise. The technique is used when the land adjacent to the sea is low in value. A decision is made to allow the land to erode and flood, creating new shoreline habitats. This process may continue over many years and natural stabilization will occur.

The earliest managed retreat in the UK was an area of 0.8 ha at Northey Island in Essex, that was flooded in 1991. This was followed by Tollesbury and Orplands in Essex, where the sea walls were breached in 1995. In the Ebro delta (Spain) coastal authorities have planned a managed retreat in response to coastal erosion (MMA 2005, Sitges, Meeting on Coastal Engineering; EUROSION project).

Cost - The main cost is generally the purchase of land to be flooded. Compensation for relocation of residents may be needed. Any other human made structure which will be engulfed by the sea may need to be safely dismantled to prevent sea pollution. In some cases, a retaining wall or bund must be constructed inland in order to protect land beyond the area to be flooded, although such structures can generally be lower than would be needed on the existing coast. Monitoring of the evolution of the flooded area is another cost. Costs may be lowest if existing defenses are left to fail naturally, but often the realignment project will be more actively managed, for example by creating an artificial breach in existing defences to allow the sea in at a particular place in a controlled fashion, or by pre-forming drainage channels for created salt-marsh.

HOLD THE LINE

Human strategies on the coast have been heavily based on a static engineered response, whereas the coast is in, or strives towards, a dynamic equilibrium (Schembri, 2009). Solid coastal structures are built and persist because they protect expensive properties or infrastructures, but they often relocate the problem downdrift or to another part of the coast. Soft options like beach nourishment, while also being temporary and needing regular replenishment, appear more acceptable, and go some way to restore the natural dynamism of the shoreline.

However in many cases there is a legacy of decisions that were made in the past which have given rise to the present threats to coastal infrastructure and which necessitate immediate shore protection. For instance, the seawall and promenade of many coastal cities in Europe represents a highly engineered use of prime seafront space, which might be preferably designated as public open space, parkland and amenities if it were available today. Such open space might also allow greater flexibility in terms of future land-use change, for instance through managed retreat, in the face of threats of erosion or inundation as a result of sea-level rise. Foredunes areas represent a natural reserve which can be called upon in the face of extreme events; building on these areas leaves little option but to undertake costly protective measures when extreme events (whether amplified by gradual global change or not) threaten. Managed retreat can comprise 'setbacks', rolling easements and other planning tools including building within a particular design life. Maintenance of those structures or soft techniques can arrive at a critical point (economically or environmental) to change adopted strategy.

- Structural or hard engineering techniques, i.e. using permanent concrete and rock constructions to "fix" the coastline and protect the assets locate behind. These techniques--seawalls, groynes, detached breakwaters, and revetments-represent a significant share of protected shoreline in Europe (more than 70%).
- Soft engineering techniques (e.g. sand nourishments), building with natural processes and relying on natural elements such as sands, dunes and vegetation to prevent erosive forces from reaching the backshore. These techniques include beach nourishment and sand dune stabilization.

MOVE SEAWARD

The futility of trying to predict future scenarios where there is a large human influence is apparent. Even future climate is to a certain extent a function of what humans choose to make of it, for example by restricting greenhouse gas emissions to control climate change. In some cases - where new areas are needed for new economic or ecological development - a move seaward strategy can be adopted. Examples from erosion include: Koge Bay (Dk) Western Scheldt estuary (Nl), Chatelaillon (F), Ebro delta (E)

There is an obvious downside to this strategy. Coastal erosion is already widespread, and there are many coasts where exceptional high tides or storm surges result in encroachment on the shore, impinging on human activity. If the sea rises, many coasts that are developed with infrastructure along or close to the shoreline will be unable to accommodate erosion. They will experience a so-called "coastal squeeze" whereby the ecological or geomorphological zones that would normally retreat landwards encounter solid structures and are squeezed out. Wetlands, salt marshes, mangroves and adjacent fresh water wetlands are particularly likely to suffer from this squeeze. An upside to the strategy is that moving seaward (and upward) can create land of high value which can bring the investment required to cope with climate change.

LIMITED INTERVENTION

Limited intervention is an action taken whereby the management only solves the problem to some extent, usually in areas of low economic significance. Measures taken using limited intervention often encourage the succession of haloseres, including salt marshes and sand dunes. This will normally result in the land behind the halosere being more sufficiently protected, as wave energy will be dissipated by the accumulated sediment and additional vegetation residing in the newly formed habitat. Although the new halosere is not strictly man-made, as many natural processes will contribute to the succession of the halosere, anthropogenic factors are partially responsible for the formation as an initial factor was needed to help start the process of succession. This must not be confused with 'accommodate' which is about property e.g. effective insurance, early warning systems and not about habitat.

CONSTRUCTION TECHNIQUES

The following is a catalogue of relevant techniques that could be employed as coastal management techniques. The costs given are very rough estimates made during 2005, based on UK Pound sterling.

SEAWALL

A seawall (or sea wall) is a form of coastal defense constructed where the sea, and associated coastal processes, impact directly upon the landforms of the coast. The purpose of a seawall is to protect areas of human habitation, conservation and leisure activities from the action of tides and waves. As a seawall is a static feature it will conflict with the dynamic nature of the coast and impede the exchange of sediment between land and sea.

The coast is generally a high-energy, dynamic environment with spatial variations occurring over a wide range of temporal scales. The shoreline is part of the coastal interface which is exposed to a wide range of erosional processes arising from fluvial, aeolian and terrestrial sources, meaning that a combination of denudational processes will work against a seawall. Given the natural forces to which seawalls are constantly subjected, maintenance (and eventually replacement) is an ongoing requirement if they are to provide an effective long term solution.

The many types of seawall in use today reflect both the varying physical forces they are designed to withstand, and location specific aspects, such as: local climate, coastal position, wave regime, and value of landform. Seawalls are classified as a hard engineering shore based structure used to provide protection and to lessen coastal erosion. However, a range of environmental problems and issues may arise from the construction of a seawall, including disrupting sediment movement and transport patterns, which are discussed in more detail below. Combined with a high construction cost, this has led to an increasing use of other soft engineering coastal management options such as beach replenishment.

Seawalls may be constructed from a variety of materials, most commonly: reinforced concrete, boulders, steel, or gabions. Additional seawall construction materials may include: vinyl, wood, aluminium, fibreglass composite, and with large biodegrable sandbags made of jute and coir. In the UK, sea wall also refers to an earthen bank used to create a polder, or a dike construction.

TRADE-OFFS

A cost benefit approach is an effective way to determine whether a seawall is appropriate and whether the benefits are worth the expense. Besides controlling erosion, consideration must be given to the effects of hardening a shoreline on natural coastal ecosystems and human property or activities. A seawall is a static feature which can conflict with the dynamic nature of the coast and impede the exchange of sediment between land and sea. The table below summarises some positive and negative effects of seawalls which can be used when comparing their effectiveness with other coastal management options, such as beach nourishment.

Advantages and disadvantages of seawalls according to Short (1999)

Advantages Disadvantages

- Long term solution in comparison to soft beach nourishment.
- Effectively minimizes loss of life in extreme events and damage to property caused by erosion.
- Can exist longer in high energy environments in comparison to 'soft' engineering methods.
- Can be used for recreation and sightseeing.
- Forms a hard and strong coastal defence.
- Very expensive to construct.
- Can cause beaches to dissipate rendering them useless for beach goers.
- Scars the very landscape that they are trying to save and provides an 'eyesore.'
- Reflected energy of waves leading to scour at base.
- Can disrupt natural shoreline processes and destroy shoreline habitats such as wetlands and intertidal beaches.
- Altered sediment transport processes can disrupt sand movement that can lead to increased erosion down drift from the structure.

Generally seawalls can be a successful way to control coastal erosion, but only if they are constructed well and out of materials which can withstand the force of ongoing wave energy. Some understanding is needed of the coastal processes and morphodynamics specific to the seawall location. Seawalls can be very helpful; they can offer a more long term solution than soft engineering options, additionally providing recreation opportunities and protection from extreme events as well as everyday erosion. Extreme natural events expose weaknesses in the performance of seawalls, and analyses of these can lead to future improvements and reassessment.

SIMULATIONS

In 2007 researchers at the University of Salerno published studies showing interactions between maritime breakwaters and waves using CAD and CFD software. In the simulations the filtration motion of the fluid within the interstices, which normally exist in a breakwater, is estimated by integrating the relevant RANS equations coupled with a random number generated turbulence model inside the voids, rather than using the classical equations for porous media.

The breakwaters were modelled, both for the actual size construction and for a physical laboratory test, by overlapping three-dimensional elements. The numerical grid was thickened in such a way to have some computational nodes along the flow paths among the breakwater's blocks.

ISSUES

Sea level rise

Sea level rise creates an issue for seawalls worldwide as it raises both the mean normal water level and the height of waves during extreme weather events, which the current seawall heights may be unable to cope with (Allan et al. 1999).

The International Panel on Climate Change (IPCC) (1997) suggested that sea level rise over the next 50 - 100 years will accelerate with a projected increase in global mean sea level of +18 cm by 2050 AD. This data is reinforced by Hannah (1990) who calculated similar statistics including a rise of between +16-19.3 cm throughout 1900-1988. This problem could be overcome by further modelling and determining the extension of height and reinforcement of current seawalls which needs to occur for safety to be ensured in both situations.

Extreme events

Extreme events also pose a problem as it is not easy for people to predict or imagine the strength of hurricane or storm induced waves compared to normal, expected wave patterns. An extreme event can dissipate hundreds of times more energy than everyday waves, and calculating structures which will stand the force of coastal storms is difficult and, often the outcome can become unaffordable.

For example, Omaha Beach seawall in New Zealand was designed to prevent erosion from everyday waves only, and when a storm in 1976 carved out 10m behind the existing seawall the whole structure was destroyed (GeoResources, 2001).

Other issues

Some further issues include: lack of long term trend data of seawall effects due to a relatively short duration of data records; modelling limitations and comparisons of different projects and their effects being invalid or unequal due to different beach types; materials; currents; and environments (Christchurch City Council, 2009).

HISTORICAL EXAMPLES

Seawall construction has existed since ancient times. In the 1st century BCE, Romans built a seawall / breakwater at Caesarea Maritima creating an artificial harbor (Sebastos Harbor). The construction used Pozzolana concrete which hardens in contact with sea water. Barges were constructed and filled with the concrete. They were floated into position and sunk. The resulting harbor / breakwater / sea wall is still in existence today - more than 2000 years later.

More recently, sea walls were constructed in 1623 in Canvey Island, UK, when great floods of the Thames estuary occurred, prompting the construction of protection for further events in this flood prone area (Council of Europe, 1999). Since then, seawall design has become more complex and intricate in response to an improvement in materials, technology and an understanding of how coastal processes operate.

This section will outline some key case studies of seawalls in chronological order and describe how they have performed in response to tsunami or ongoing natural processes and how effective they were in these situations. Analysing the successes and shortcomings of seawalls during severe natural events allows their weaknesses to be exposed, and areas become visible for future improvement.

Pondicherry

On December 26, 2004, towering waves of the 2004 Indian Ocean earthquake tsunami crashed against India's south-eastern coastline killing thousands. However, the former French colonial enclave of Pondicherry (now Pondicherry) escaped unscathed. This was primarily due to French engineers who had constructed (and maintained) a massive stone seawall during the time which the city was a French colony. This 300 year old seawall effectively kept Pondicherry's historic centre dry even though tsunami waves drove water 24 feet above the normal high-tide mark.

The barrier was initially completed in 1735 and over the years, the French continued to fortify the wall, piling huge boulders along its 1.25 mile (2 km) coastline to stop erosion from the waves pounding the harbour. At its highest, the barrier running along the water's edge reaches about 27 feet above sea level. The boulders, some weighing up to a ton, are weathered black and brown. The sea wall is inspected every year and whenever gaps appear or the stones sink into the sand, the government adds more boulders to keep it strong (Allsop, 2002). The Union Territory of Pondicherry recorded some 600 deaths from the huge tsunami waves that struck India's coast after the mammoth underwater earthquake (which measured 9.0 on the moment magnitude scale) off Indonesia, but most of those killed were fishermen who lived in villages beyond the artificial barrier which reinforces the effectiveness of seawalls.

Vancouver

The Vancouver Seawall is a stone seawall constructed around the perimeter of Stanley Park in Vancouver. The seawall was constructed initially as waves created by ships passing through the First Narrows were eroding the area between Prospect Point and Brockton Point. The Vancouver Seawall also exemplifies how seawalls can be utilised and valued for recreational activities and coastal sightseeing. A pedestrian, cycling and roller blading pathway exists on the seawall and has been extended far outside the parameters of Stanley Park. Construction of the seawall began in 1917, and since then this pathway has become one of the most used features of the park by both locals and tourists and now extends 22 km in total (Belyea & Ross, 1992). The construction of the seawall also provided employment for relief workers during the Great Depression and seamen from HMCS Discovery on Deadman's Island who were facing punishment detail in the 1950s (Steele, 1985).

Overall, the Vancouver Seawall is a prime example of how seawalls can simultaneously provide shoreline protection and a source of recreation which enhances human enjoyment of the coastal environment. It also illustrates that although shoreline erosion is a natural process, human activities, interactions with the coast and poorly planned shoreline development projects can accelerate natural erosion rates.

Japan

At least 43 percent of Japan's 29,751 km kilometre coastline is lined with concrete seawalls or other structures designed to protect the country against high waves, typhoons or even tsunamis (New York Times, 2011). When a Tsunami struck in 2011 following a magnitude 9 offshore earthquake, the seawalls in most areas were overwhelmed. In Kamaishi, 4-metre waves surmounted the seawall -the world's largest, erected a few years ago in the city's harbour at a depth of 63 metres, a length of 2 kilometres and a cost of $1.5 billion - and eventually submerged the city centre (Musubi, 2011).

The risks of dependence on seawalls was most evident in the crisis at the Dai-ichi and Dai-ni nuclear power plants, both located along the coast close to the earthquake zone, as the tsunami washed over walls that were supposed to protect the plants. Arguably, the additional defence provided by the seawalls presented an extra margin of time for citizens to evacuate and also stopped some of the full force of energy which would have caused the wave to climb higher in the backs of coastal valleys. In contrast, the seawalls also acted in a negative way to trap water and delay its retreat.

The failure of the world's largest seawall, which cost $1.5 billion to construct, shows that building stronger sea walls to protect larger areas would have been too costly to be effective. In the case of the ongoing crisis at the nuclear power plants, higher and stronger sea walls should have been built if power plants were to be built at that site.

Fundamentally, the devastation in coastal areas and a final death toll predicted to exceed 10,000 could push Japan to redesign its seawalls or consider more effective alternative methods of coastal protection for extreme events. Such hardened coastlines can also provide a false sense of security to property owners and local residents as evident in this situation (Msubi, 2011).

BREAKWATER (STRUCTURE)

Breakwaters are structures constructed on coasts as part of coastal defense or to protect an anchorage from the effects of both weather and longshore drift.

PURPOSES OF BREAKWATERS

Offshore breakwaters, also called bulkhead, reduce the intensity of wave action in inshore waters and thereby reduce coastal erosion or provide safe harborage. Breakwaters may also be small structures designed to protect a gently sloping beach and placed one to three hundred feet offshore in relatively shallow water.

An anchorage is only safe if ships anchored there are protected from the force of high winds and powerful waves by some large underwater barrier which they can shelter behind. Natural harbours are formed by such barriers as headlands or reefs. Artificial harbors can be created with the help of breakwaters. Mobile harbours, such as the D-Day Mulberry harbours, were floated into position and acted as breakwaters. Some natural harbours, such as those in Plymouth Sound, Portland Harbour and Cherbourg, have been enhanced or extended by breakwaters made of rock.

UNINTENDED CONSEQUENCES

The dissipation of energy and relative calm water created in the lee of the breakwaters often encourage accretion of sediment (as per the design of the breakwater scheme). However this can lead to excessive salient build up, leading to tombolo formation reducing longshore drift shoreward of the breakwaters (Sea Palling, UK). This trapping of sediment can cause adverse effects down drift of the breakwaters leading to beach sediment starvation and increased erosion. This may then lead to further engineering protection being needed down drift of the breakwater development. Breakwaters are subject to damage, and overtopping in severe storms events.

CONSTRUCTION

Breakwaters can be constructed with one end linked to the shore, in which case they are usually classified as sea walls; otherwise they are positioned offshore from as little as 100m up to 300-600m from the original shoreline. There are two main types of offshore breakwater, single and multiple; single as the name suggests means the breakwater consists of one unbroken barrier, which multiple breakwaters (in numbers anywhere from 2-20) are positioned with gaps in between (50-300m).

Length of gap is largely governed by the interacting wavelengths. Breakwaters may be either fixed or floating, and impermeable or permeable to allow sediment transfer shoreward of the structures, the choice depending tidal range and water depth. They usually consist of large pieces of rock (granite) weighting up to 16 tonnes each or rubble-mound. Their design is influenced by the angle of wave approach and other environmental parameters. Breakwater construction can be either parallel or perpendicular to the coast, depending on the shoreline requirements.

TYPES OF BREAKWATER STRUCTURES

A breakwater structure is designed to absorb the energy of the waves that hit it, either by using mass (e.g., with caissons), or by using a revetment slope (e.g., with rock or concrete armour units). In Coastal Engineering, a revetment is a land backed structure whilst a breakwater is a sea backed structure (i.e., water on both sides).

Caisson breakwaters typically have vertical sides and are usually erected where it is desirable to berth one or more vessels on the inner face of the breakwater. They use the mass of the caisson and the fill within it to resist the overturning forces applied by waves hitting them. They are relatively expensive to construct in shallow water, but in deeper sites they can offer a significant saving over revetment breakwaters. Rubble mound breakwaters use structural voids to dissipate the wave energy. Rock or concrete armour units on the outside of the structure absorb most of the energy, while gravels or sands prevent the wave energy's continuing through the breakwater core. The slopes of the revetment are typically between 1:1 and 1:2, depending upon the materials used. In shallow water, revetment breakwaters are usually relatively inexpensive. As water depth increases, the material requirements, and hence costs, increase significantly.

ADVANCED NUMERICAL STUDY

The Maritime Engineering Division of the University of Salerno (MEDUS) developed a new procedure for studying in greater detail the interactions between maritime breakwaters (submerged or emerged) and the waves that hit them by making integrated use of CAD and CFD software. In the numerical simulations, the filtration motion of the fluid within the interstices, which normally exist in a breakwater, is estimated by integrating the RANS equations, coupled with a RNG turbulence model inside the voids, instead of using classical equations for porous media.

The breakwaters were modelled, in analogy to full size construction or physical laboratory tests, by overlapping three-dimensional elements and having the numerical grid thickened in order to have some computational nodes along the flow paths among the breakwater's blocks.

CLIFF STABILIZATION

Cliff stabilization is a coastal management erosion control technique. This is most suitable for softer or less stable cliffs. Generally speaking, the cliffs are stabilised through dewatering (drainage of excess rainwater to reduce water-logging) or anchoring (the use of terracing, planting, or wiring to hold cliffs in place).

ENTRANCE TRAINING WALLS

Rock or concrete walls built to constrain a river or creek discharging across a sandy coastline. The walls help to stabilise and deepen the channel which benefits navigation, flood management, river erosion and water quality but can cause coastal erosion due to the interruption of longshore drift. One solution is the installation of a sand bypassing system to pump sand under and around the entrance training walls.

Cost - Expensive - Gold Coast Seaway was a A$50M project in the 1980s and the adjacent sand bypassing project costs A$3M per year to pump 500,000 cubic meters of sand across the trained entrance.

FLOODGATES

Storm surge barriers, or floodgates, were introduced after the North Sea Flood of 1953 and are a prophylactic method to prevent damage from storm surges or any other type of natural disaster that could harm the area they "protect". They are habitually open and allow free passage, but close when the land is under threat of a storm surge. The Thames Barrier is an example of such a structure.

BEACH NOURISHMENT

Beach nourishment- also referred to as beach replenishment or sand replenishment -describes a process by which sediment (usually sand) lost through longshore drift or erosion is replaced from sources outside of the eroding beach. A wider beach can reduce storm damage to coastal structures by dissipating energy across the surf zone, protecting upland structures and infrastructure from storm surges, tsunamis and unusually high tides. Beach nourishment is typically part of a larger coastal defense scheme. Nourishment is typically a repetitive process, since it does not remove the physical forces that cause erosion, but simply mitigates their effects.

The first nourishment project in the U.S. was at Coney Island, New York in 1922-23 and is now a common shore protection measure utilized by public and private entities. Nourishment is one of three commonly accepted methods for protecting shorelines. The structural alternative involves constructing a seawall, revetment, groin or breakwater. Alternatively, with "managed retreat" the shoreline is left to erode, while relocating buildings and infrastructure further inland. Nourishment gained popularity because it preserved beach resources and avoided the negative effects of hard structures. Instead, nourishment creates a "soft" (i.e., non-permanent) structure by creating a larger sand reservoir, pushing the shoreline seaward.

CAUSES OF EROSION

Beaches can erode both naturally and due to impact of humans. Erosion is a natural response to storm activity. During storms, sand from the visible beach submerges to form storm bars that protect the beach. Submersion is only part of the cycle. During calm weather smaller waves return sand from the storm bar to the visible beach surface in a process called accretion. The term erosion conjures visions of environmental damage so the term submersion often replaces it in describing a healthy sandy beach. Continental drift erodes coastlines naturally. Ocean currents can change, disrupting the submersion/accretion cycle.

Some beaches do not have enough sand available to coastal processes to respond naturally to storms. When there is not enough sand left available on a beach, then there is no recovery of the beach following storms. Many areas of high erosion are due to human activities. Reasons can include seawalls locking up sand dunes, coastal structures like ports and harbors that prevent longshore drift, dams and other river management structures. These activities interfere with the natural sediment flows either through dam construction (thereby reducing riverine sediment sources) or construction of littoral barriers such as jetties, or by deepening of inlets; thus preventing longshore transport of sediment across these channels.

VISIBLE AND SUBMERGED SAND

The distinction between total sand in a beach and the proportion of the sand above the waterline (submersion fraction) critically impacts beach nourishment. Two beaches with the same amount of visible sand may look much different under water. An eroded beach with substantial submerged sand surrounding it may recover without nourishment. Nourishing a beach that has little submerged sand requires addressing the reason that the submerged sand is missing. Otherwise, the same forces that stripped the submerged sand once are likely to do so again. The amount of submerged sand eroded is typically much greater than the amount of missing sand on shore. Replacing only the visible sand is insufficient without replacing the sand that once supported the accretion part of the process. In that circumstance, the beach is unstable and the visible sand quickly erodes. If human activity is a major cause of the erosion, mitigating that activity may be more cost effective over both short and long term periods than nourishment.

REQUIREMENTS FOR EFFECTIVE NOURISHMENT

Sediment texture (grain size and sorting) is critical for success. Sand fill must be compatible with native beach sand. In some cases, beaches have been nourished using a finer sand than the original. Thermoluminescence monitoring reveals that storms can erode such beaches far more quickly than the natural beach. This was observed at the Waikiki nourishment project in Hawaii.

Profile Nourishment

Beach Profile Nourishment describes programs that nourish the full beach profile. In this instance, "profile" means the slope of the uneroded beach from above the water to well out to sea, not just the visible portion. The Gold Coast profile nourishment program placed 75% of its total sand volume below low water level. Some coastal authorities overnourish the below water beach (aka "nearshore nourishment") so that over time the natural beach increases in size. These approaches do not permanently protect beaches eroded by human activity. Doing so still requires mitigating that activity.

ENVIRONMENTAL ISSUES

Beach nourishment has significant impacts on local ecosystems. Nourishment may cause direct mortality to sessile organisms in the target area by burying them in the new sand. Seafloor habitat in both source and target areas are disrupted, e.g., when sand is deposited on coral reefs or when deposited sand hardens. Imported sand may differ in character (chemical makeup, grain size, non-native species) from that of the target environment. Reduced light availability, affecting nearby reefs and submerged aquatic vegetation. Imported sand may contain material toxic to local species. Removing material from near-shore environments may destabilize the shoreline, in part by steepening its submerged slope. Related attempts to reduce future erosion may provide a false sense of security that increases development pressure.

Sea turtles

Newly-deposited sand can harden and complicate nest-digging for turtles.

However, nourishment can provide more/better habitat for them, as well as for sea birds and beach flora. Florida addressed the concern that dredge pipes would suck turtles into the pumps by adding a special grill to the dredge pipes that ensured that turtles were not sucked in with the sand.

ALTERNATIVES/COMPLEMENTS TO NOURISHMENT

Nourishment is not the only technique used to address eroding beaches. Others can be used singly or in combination with nourishment, driven by economic, environmental, and political considerations. Human activities such as dam construction can interfere with natural sediment flows (thereby reducing riverine sediment sources.) Construction of littoral barriers such as jetties and deepening of inlets can prevent longshore sediment transport.

Structural

The structural approach attempts to prevent erosion. Armoring involves building revetments, seawalls, detached breakwaters, groins, etc. Structures that run parallel to the shore (seawalls or revetments) prevent erosion. While this protects structures, it doesn't not protect the beach which is outside the wall. The beach disappears over a period that ranges from months to decades. Groynes and breakwaters that run perpendicular to the shore protect the shore from erosion. Filling a breakwater with imported sand can stop the breakwater from trapping sand from the littoral stream (the ocean running along the shore.) Otherwise the breakwater may deprive downstream beaches of sand and accelerate erosion there. Armoring may restrict beach/ocean access, enhance erosion of adjacent shorelines, and requires long-term maintenance.

Managed retreat

A second option is to move structures and infrastructure inland as the shoreline erodes. Retreat is more often chosen in areas of rapid erosion and in the presence of little or obsolete development.

Recruitment

Appropriately constructed and sited fences can capture blowing sand, building/restoring sand dunes, and progressively protecting the beach from the wind, and the shore from blowing sand.

Beach drainage

All beaches lose and gain sand depending on the tides, precipitation, wind, waves etc. Wet beaches tend to lose sand. Waves infiltrate dry beaches easily and deposit sandy sediment. Generally a beach is wet during falling tide, because the sea sinks faster than the beach drains. As a result most erosion happens during falling tide. Beach drainage (beach dewatering) using Pressure Equalizing Modules (PEMs) allow the beach to drain more effectively during falling tide. Fewer hours of wet beach translate to less erosion. Permeable PEM tubes inserted vertically into the foreshore connect the different layers of groundwater. The groundwater enters the PEM tube allowing gravity to conduct it to a coarser sand layer, where it can drain more quickly.

The PEM modules are placed in a row from the dune to the mean low waterline. Distance between rows is typically 300 feet (91 m) but this is project-specific. PEM systems come in different sizes. Modules connect layers with varying hydraulic conductivity. Air/water can enter and equalize pressure. PEMs are minimially invasive, typically covering approximately 0.00005% of the beach.

The effects are local, and bring no known harm to flora or fauna, including nesting sea turtles. The tubes are below the beach surface, with no visible impact. Installation takes little time. Costs are low, because PEMs require few materials, and no energy-intensive dredging. PEM installations have been installed on beaches in Denmark, Sweden, Malaysia, and Florida, USA. The effectiveness of beach dewatering, however, is debatable and has not been proven convincingly on life-size beaches.

COSTS

Nourishment is typically a repetitive process, since nourishment does not remove the physical forces that cause erosion; it simply mitigates their effects. A benign environment increases the interval between nourishment projects, reducing costs. Conversely, high erosion rates may render nourishment financially impractical.

In many coastal areas, the economic impacts of a wide beach can be substantial. The 10 miles (16 km)-long shoreline fronting Miami Beach, Florida was replenished over the period 1976-1981. The project cost approximately $64,000,000 and revitalized the area's economy. Prior to nourishment, in many places the beach was too narrow to walk along, especially during high tide.

HISTORY

The first nourishment project in the U.S. was constructed at Coney Island, New York in 1922-1923.

NOURISHMENT PROJECTS

The setting of a beach nourishment project is key to design and potential performance. Possible settings include a long straight beach, an inlet that may be either natural or modified, and a pocket beach. Rocky or seawalled shorelines, that otherwise have no sediment, present unique problems.

Cancun, Mexico

Federal and state governments in Mexico have invested about $71 million ($957 million pesos) throughout the state of Quintana Roo in restoring the beaches along Cancun, Playa del Carmen, and Cozumel. Hurricane Wilma hit the beaches of Cancun and the Riviera Maya in 2005. The initial nourishment project was unsuccessful, leading to a second round that began in September 2009, and was scheduled to complete in early 2010.

The project designers and the government committed to invest in beach maintenance to address future erosion. Project designers considered factors such as the time of year and sand characteristics such as density. Restoration in Cancun was expected to deliver 1.3 billion US gallons (4,900,000 m) of sand to replenish 450 meters (1,480 ft) of coastline. This time, the beach is promised to last at least 10 years.

Northern Gold Coast, Queensland, Australia

Gold Coast beaches in Queensland, Australia have experienced periods of severe erosion. In 1967 a series of 11 cyclones removed most of the sand from Gold Coast beaches. The Government of Queensland engaged engineers from Delft University in the Netherlands to advise them. The 1971 Delft Report outlined a series of works for Gold Coast Beaches, including beach nourishment and an artificial reef. By 2005 most of the recommendations had been implemented.

The Northern Gold Coast Beach Protection Strategy (NGCBPS) was a A$10 million investment. NGCBPS was developed between 1992 and 1999 and the works were completed between 1999 and 2003. The project included dredging 3,500,000 cubic metres (4,600,000 cu yd) of compatible sand from the Gold Coast Broadwater and delivering it through a pipeline to nourish 5 kilometers (3.1 mi) between of beach between Surfers Paradise and Main Beach.

The new sand was stabilized by an artificial reef constructed at Narrowneck out of huge geotextile sand bags. The new reef was designed to improve wave conditions for surfing. A key monitoring program for the NGCBPS is the ARGUS coastal camera system operated by the University of New South Wales.

The cost/benefit ratio for NGCBPS was conservatively estimated at 75:1 for a A$10 million investment into beach replenishment. The benefits were estimated from a model of lost visitor nights in hotels following previous erosion events. NGCBPS so improved beach health that recovery following minor and moderate storms occurred within weeks. Additional unquantified benefits included lifestyle benefits for residents, additional public open space and improved fishing, diving and surfing conditions.

Netherlands

More than one-quarter of the Netherlands is below sea level and about 81% of the coast consists of sand dune or beach. The shoreline is closely monitored by yearly recording of the cross section at points 250 meters (820 ft) apart, to ensure adequate protection. Where long-term erosion is identified, beach nourishment using high-capacity suction dredgers is deployed.

Hawaii

Hawaii planned to replenish Waikiki beach in 2010. Budgeted at $2.5 million, the project covered 1,700 feet (520 m) in an attempt to return the beach to its 1985 width. Prior opponents supported this project, because the sand was to come from nearby shoals, reopening a blocked channel and leaving the overall local sand volume unchanged, while closely matching the "new" sand to existing materials. The project planned to apply up to 24,000 cubic yards (18,000 m) of sand from deposits located 1,500 to 3,000 feet (460 to 910 m) offshore at a depth of 10 to 20 feet (3.0 to 6.1 m). The project was larger than the prior recycling effort in 2006-07, which moved 10,000 cubic yards (7,600 m).

Maui, Hawaii illustrated the complexities of even small-scale nourishment projects. A project at Sugar Cove transported upland sand to the beach. The sand allegedly was finer than the original sand and contained excess silt that enveloped coral heads, smothering the coral and killing small animals that lived in and around it.

As in other projects, on-shore sand availability was limited, forcing consideration of more expensive offshore sources. A second project, along Stable Road, that attempted to slow rather than halt erosion, was stopped halfway toward its goal of adding 10,000 cubic yards (7,600 m) of sand. The beaches had been retreating at a "comparatively fast rate" for half a century. The restoration was complicated by the presence of old seawalls, groins, piles of rocks and other structures.

This project used sand-filled Geotube groins placed that were originally to remain in place for up to 3 years. The pipeline was anchored by concrete blocks attahced by fiber straps. A video showed the blocks bouncing off the coral in the current, killing whatever they touched. In places the straps broke, allowing the pipe to move across the reef, "planing it down". Bad weather exacerbated the damaging movement and led to project shutdown. The smooth, cylindrical Geotubes could be difficult to climb over before they were covered by sand.

Supporters claimed that 2010's seasonal summer erosion was less than in prior years, although the beach was narrower after the restoration ended than in 2008. Authorities were studying whether to require the project to remove the groins immediately. Potential alternatives to Geotubes for moving sand included floating dredges and/or trucking in sand dredged offshore. A final consideration was sea level rise and that Maui was sinking under its own weight. Both islands surround massive mountains (Haleakala, Mauna Loa, and Mauna Kea) and were expanding a giant dimple in the ocean floor, some 30,000 feet (9,100 m) below the mountain summits.

Hong Kong

The beach in Gold Coast was built as an artificial beach in the 1990s with HK$60m. Sands have to be supplied periodically, especially after typhoons, to keep the beach viable.

MEASURING PROJECT IMPACT

Nourishment projects usually involve physical, environmental and economic objectives. Typical physical measures include dry beach width, remaining post-storm sand volume, post-storm damage avoidance assessments and aqueous sand volume. Environmental measures include marine life distribution, habitat and population counts. Economic impacts include recreation, tourism, flood and "disaster" prevention. Techniques for incorporating nourishment projects into flood insurance costs and disaster assistance remain controversial.

The ability to predict the performance of a beach nourishment project is best for a project constructed on a long, straight shoreline without the complications of inlets or engineered structures. In addition, predictability is better for overall performance, e.g., average shoreline change, rather than shoreline change at a specific location. Nourishment can affect eligibility in the National Flood Insurance Program and federal disaster assistance.

Sand dune stabilization

Vegetation can be used to encourage dune growth by trapping and stabilising blown sand. Cost - est. of £1.1 million per annum

Beach drainage

Beach drainage or beach face dewatering lowers the water table locally beneath the beach face. This causes accretion of sand above the drainage system. Grant (1946) - the elevation of the beach watertable had an important bearing on deposition and erosion across the foreshore. A high watertable coincided with periods of accelerated beach erosion, and conversely, a low watertable coincided with pronounced aggradation of the foreshore A lower watertable (unsaturated beach face) facilitates deposition by reducing flow velocities during backwash and prolonging laminar flow. In contrast, a high watertable results in condition favoring beach erosion. With the beach in a saturated state, Grant proposed that backwash velocity is accelerated by the addition of groundwater seepage out of the beach within the effluent zone.

Turner and Leatherman (1997) moving from the origins and development of the dewatering concept to field and laboratory studies available at the time of writing concluded that there was too little evidence for being convinced that the systems had a positive effect. None of the case studies provide full scientific evidence of indisputable positive results regarding beach stabilisation although in some cases an overall positive performance was reported. In many cases no adequate long-term monitoring was undertaken at a frequency high enough to discriminate the response to high energy erosive events.

A useful side effect of the system is that the collected seawater is very pure because of the sand filtration effect. It may be discharged back to sea but can also be used to oxygenate stagnant inland lagoons /marinas or used as feed for heat pumps, desalination plants, land-based aquaculture, aquariums or seawater swimming pools. Beach drainage systems have been installed in many locations around the world to halt and reverse erosion trends in sand beaches. Twenty four beach drainage systems have been installed since 1981 in Denmark, USA, UK, Japan, Spain, Sweden, France, Italy and Malaysia.

Costs

The costs of installation and operation per meter of shoreline protection will vary due to

- system length (non-linear cost elements)
- pump flow rates (sand permeability, power costs)
- soil conditions (presence of rock or impermeable strata)
- discharge arrangement /filtered seawater utilization
- drainage design, materials selection & installation methods
- geographical considerations (location logistics)
- regional economic considerations (local capabilities /costs)
- study requirements /consent process.

The costs associated with a beach drainage system are generally considerably lower than hard engineered structures. They also compare very favorably with beach nourishment projects, particularly when long-term project economics are considered (nourishment projects often have a limited life or a program of re-nourishment).

MONITORING COASTAL ZONES

Coastal zone managers are faced with difficult and complex choices about how best to reduce property damage in the shorelines. One of the problems they face is error and uncertainty in the information available to them on the processes that cause erosion of beaches. Video-based monitoring lets collect data continuously at low cost and produce analyses of shoreline processes over a wide range of averaging intervals.

EVENT WARNING SYSTEMS

Event warning systems, such as tsunami warnings and storm surge warnings, can be used to minimize the human impact of catastrophic events that cause coastal erosion. Storm surge warnings can also be used to determine when to close floodgates to reduce the physical impact of such events. Wireless sensor networks can be deployed quickly to set up a coastal erosion monitoring system, and scaled accordingly.

SHORELINE MAPPING

Defining the shoreline is a difficult task due to the dynamic nature of the coast and the intended application of the shoreline (Graham et al. 2003; Boak & Turner 2005). Given this idea the shoreline must therefore be considered in a temporal sense whereby the scale is dependent on the context of the investigation (Boak & Turner 2005). The following definition of the coast and shoreline is most commonly employed for the purposes of shoreline mapping. The coast comprises the interface between land and sea, and the shoreline is represented by the margin between the two (Woodroffe, 2002). Due to the dynamic nature of the shoreline coastal investigators adopt the use of shoreline indicators to represent the true shoreline position (Boak & Turner 2005).

SHORELINE INDICATOR

The choice of shoreline indicator is a primary consideration in shoreline mapping. According to Leatherman (2003) it is important that indicators are easily identified in the field and on aerial photography. Shoreline indicators may be physical beach morphological features such as the berm crest, scarp edge, vegetation line, dune toe, dune crest and cliff or the bluff crest and toe. Alternatively, non-morphological features may also be used. These indicators are based on water level including the high water line, mean high water line, wet/dry boundary, and the physical water line (Pajak & Leatherman 2000). The high water line (HWL), defined as the wet/dry line is the most commonly used shoreline indicator because it is visible in the field, and can be interpreted on both colour and grey scale aerial photographs (Leatherman, 2003; Crowell et al. 1991). The HWL represents the landward extent of the most recent high tide and is characterised by a change in sand colour due to repeated, periodic inundation by high tides. The HWL is portrayed on aerial photographs by the most landward change in colour or grey tone (Boak & Turner 2005).

IMPORTANCE AND APPLICATION

The location of the shoreline and its changing position over time is of fundamental importance to coastal scientists, engineers and managers (Boak & Turner 2005; Pajack & Leatherman 2002). Present day shoreline monitoring campaigns provide information about historic shoreline location and movement, and about predictions of future change (Appeaning Addo et al. 2008). More specifically the position of the shoreline in the past, at present and where it is predicted to be in the future is useful for in the design of coastal protection, to calibrate and verify numerical models to assess sea level rise, map hazard zones and formulate policies to regulate coastal development. Accurate and consistent delineation of the shoreline is integral to all of these tasks. The location of the shoreline also provides information regarding shoreline reorientation adjacent to structures, beach width, volume and rates of historical change (Boak & Turner 2005; Pajack & Leatherman 2002).

DATA SOURCES

A variety of data sources are available for examining shoreline position however, the availability of historical data is limited at many coastal sites and so the choice of data source is largely limited to what is available for the site at a given time (Boak & Turner 2005). Shoreline mapping techniques applied to data sources have moved towards automation in association with technological advances and the need to reduce uncertainty.

Although these changes have resulted in improvement in coastal data processing and storage capabilities, the frequent change in technology has prevented the emergence of one standard method of shoreline mapping. This has occurred because each data source and associated method have their own unique capabilities and shortcomings (Moore 2000). A number of the data sources used for shoreline mapping and their associated advantages and disadvantages are discussed below.

Historical maps

In the event that a study requires the shoreline position to be mapped before the development of aerial photographs, or if the location has poor photograph coverage it is necessary to employ historical maps in order to detail shoreline position (Moore 2000). The main advantage and reason for using historical maps is that they are able to provide a historic record that is not available from other data sources. Many potential errors however are associated with historical coastal maps and charts.

Such errors may be associated with scale, datum changes, distortions from uneven shrinkage, stretching, creases, tears and folds, different surveying standards, different publication standards, and projection errors (Boak & Turner 2005). The severity of these errors depends on the accuracy standards met by each map and the physical changes that have occurred since the publication of the map (Anders & Byrnes 1991). The oldest reliable source of shoreline data in the United States dates back to the early-to-mid-19th century and is the U.S Coast and Geodetic Survey/National Ocean Service T-sheets (Morton 1991). In the United Kingdom, many maps and charts were deemed to be inaccurate until around 1750. The founding of the Ordnance Survey in 1791 has since improved the accuracy of the mapping.

Aerial photographs

Aerial photographs have been used since the 1920s to provide topographical information about an area. They are therefore a good database for compilation of historical shoreline change maps. Aerial photographs are the most commonly used data source in shoreline mapping because many coastal areas have extensive aerial photo coverage therefore providing a valuable record of shoreline position (Moore 2000). In general, aerial photographs provide good spatial coverage of the coast however temporal coverage is very much site specific depending on the flight path of the aeroplane. A second disadvantage associated with aerial photography is that the interpretation of the shoreline position is subjective given the dynamic nature of the coastal environment. This combined with various distortions inherent in aerial photographs can lead to significant error levels (Moore 2000). The minimisation of further errors is discussed below.

Object space displacements

Conditions outside of the camera can cause objects in an image to be displaced from their true ground position. Such conditions may include ground relief, camera tilt and atmospheric refraction. Relief displacement is prominent when photographing a variety of elevations. This situation causes objects above ground level to be displaced outward from the centre of the photograph and objects below ground level to be displaced toward the centre of the image.

The severity of the displacement is affected negatively with decreases in flight altitude and as radial distance from the centre of the photograph increases. This distortion can be minimised by photographing numerous swaths and creating a mosaic of the images. This technique will create a focus for the centre of each photograph where distortion is minimised. It is important to note that this error is not common in shoreline mapping is the relief is fairly constant. It is however important to consider when mapping cliffs (Moore 2000).

Ideally aerial photographs are taken so the optical axis of the camera is perfectly perpendicular to the ground surface thereby creating a vertical photograph. Unfortunately this is not often the case and virtually all aerial photographs experience tilt whereby up to 3° is not uncommon (Camfield et al. 1996). In this situation the scale of the image will be larger on the upward side of the tilt axis and smaller on the downward side. Moore, (2000) notes that many coastal researchers have not realised the severity of this error and therefore do not consider it in their methods.

Radial lens distortion

Lens distortion varies as a function of radial distance from the iso-centre of the photograph meaning that the centre of the image is relatively distortion free, but as the angle of view increases the distortion becomes more prominent. This is a significant source of error in earlier aerial photography but as technology has increased and camera lens have become more refined it has become less of an issue with later photographs.

Such a distortion is impossible to correct for without knowing the make and model of the lens used to capture the image. However if overlapping images have been acquired one can digitize the centre portions of the aerial photographs.

Delineation of the shoreline

The dynamic nature of the coast has meant that accurate mapping of an instantaneous shoreline position has been associated with significant uncertainty. This uncertainty arises because at any given time the position of the shoreline is influenced by the short-term effect of the tide and a wide variety of long term effects such as relative sea-level rise and along shore littoral sediment movement.

Not only does this affect the accuracy of computed historic shoreline position but also any predicted future positions (Appeaning Addo et al. 2008). As mentioned earlier the HWL is most commonly used as a shoreline indicator. This can usually be seen as a significant tonal change on aerial photographs. There are however many errors associated with using the wet/dry line as a proxy for the HWL and shoreline. The errors of largest concern are the short term migration of the wet/dry line, interpretation of the wet/dry line on a photograph and measurement of the interpreted line position (Leatherman 2003; Moore 2000).

Systematic errors such as the migration of the wet/dry line may arise from tidal and seasonal changes. Storm-induced erosion is another factor which may cause the wet/dry line to migrate landward. Field investigations have shown that these changes can be minimised by using only summertime data (Moore 2000; Leatherman 2003). Furthermore, the error bar can be significantly reduced by using the longest record of reliable data to calculate erosion rates (Leatherman 2003). Finally it is important to note that errors may arise due to the difficulty of measuring a single line on a photograph. For example where the pen line is 0.13 mm thick this translates to an error of ±2.6 m on a 1:20000 scale photograph.

Beach profiling surveys

Beach profiling surveys are typically repeated at regular intervals along the coast in order to measure short-term (daily to annual) variations in shoreline position and beach volume. (Smith & Zarillo 1990). Beach profiling is a very accurate source of information however measurements are generally subject to the limitations of conventional surveying techniques. Shoreline data derived from beach profiling is often spatially and temporally limited due to the high cost associated with such a labour intensive activity.

Shorelines are generally derived by interpolating between a series of discrete beach profiles. It is important to note however that the distance between the profiles is usually quite large and so the accuracy of the interpolating becomes compromised. In contrast to aerial photographs, survey data is limited to smaller lengths of shoreline generally less than ten kilometres (Boak & Turner 2005). Beach profiling data is commonly available in from regional councils in New Zealand such as those compiled by the Hawkes Bay Regional Council.

Remote sensing

Technological advancement over the last decade has led to the development of a range of airborne, satellite and land based remote sensing techniques (Smith & Zarillo 1990). Some of the remotely sensed data sources are listed below:

- Multispectral and hyperspectral imaging
- Microwave sensors
- Global positioning system (GPS)
- Airborne light detection and ranging technology (LIDAR)

Remote sensing techniques are attractive as they are cost effective, reduce manual error and remove the subjective approach of conventional field techniques (Maiti et al. 2009).Remote sensing is a relatively new concept and so extensive historical observations are unavailable. Given this idea, it is important that coastal morphology observations are quantified by coupling remotely sensed data with other sources of information detailing historic shoreline position from archived sources (Appeaning Addo et al. 2008).

Video analysis

Video analysis provides quantitative, cost-effective, continuous and long-term monitoring beaches (Turner et al. 2004). The advancement of coastal video systems over the past 15 years has resulted in the extraction of large amounts of geophysical data from images. Such data includes that about coastal morphology, surface currents and wave parameters. The main advantage of video analysis lies in the ability to reliably quantify these parameters with high resolution and coverage in both space and time. This in particular highlights their potential importance as an effective coastal monitoring system and an aid to coastal zone management (Van Koningsveld et al. 2007).

Interesting case studies have been carried out using video analysis. Turner et al. (2004) used a video-based ARGUS coastal imaging system [4] to monitor and quantify the regional-scale coastal response to sand nourishment and construction of the world-first Gold Coast artificial (surfing) reef in Australia. In addition, Smit et al. (2007) demonstrated the added value of high resolution video observations for making short-term predictions of near shore hydrodynamic and morphological processes, at temporal scales of meters to kilometres and days to seasons.

INTEGRATED COASTAL ZONE MANAGEMENT

Integrated coastal zone management (ICZM) or Integrated coastal management (ICM) is a process for the management of the coast using an integrated approach, regarding all aspects of the coastal zone, including geographical and political boundaries, in an attempt to achieve sustainability. This concept was born in 1992 during the Earth Summit of Rio de Janeiro.

The specifics regarding ICZM is set out in the proceedings of the summit within Agenda 21, Chapter 17.

The European Commission defines the ICZM as follows: ICZM is a dynamic, multidisciplinary and iterative process to promote sustainable management of coastal zones. It covers the full cycle of information collection, planning (in its broadest sense), decision making, management and monitoring of implementation. ICZM uses the informed participation and cooperation of all stakeholders to assess the societal goals in a given coastal area, and to take actions towards meeting these objectives. ICZM seeks, over the long-term, to balance environmental, economic, social, cultural and recreational objectives, all within the limits set by natural dynamics. 'Integrated' in ICZM refers to the integration of objectives and also to the integration of the many instruments needed to meet these objectives. It means integration of all relevant policy areas, sectors, and levels of administration. It means integration of the terrestrial and marine components of the target territory, in both time and space.

To further understand the idea of ICZM several aspects can be defined and further explained. The coastal zone, the concept of sustainability and the term integration all within a coastal management context can be individually defined, while the expectations and framework of ICZM can be further explained. This entry uses the example of the New Zealand national framework to illustrate ICZM.

DEFINING THE COASTAL ZONE

Defining the Coastal zone is of particular importance to the idea of ICZM. But the fuzziness of borders due to the dynamic nature of the coast makes it difficult to clearly define. Most simply the coast can be thought of as an area of interaction between the land and the ocean. Ketchum (1972) defined the area as: The band of dry land and adjacent ocean space (water and submerged land) in which terrestrial processes and land uses directly affect oceanic processes and uses, and vice versa.

Issues arise with the diversity of features present on the coast and the spatial scales of the interacting systems. Coasts being dynamic in nature are influenced differently all around the world. Influences such as river systems, may reach far inland increasing the complexity and scale of the zone. These issues make it difficult to clearly identify hinterlands and subscribe any subsequent management.

Whilst acknowledging a physical coastal zone, the inclusion of ecosystems, resources and human activity within the zone is important. It is the human activities that warrant management. These activities are responsible for disrupting the natural coastal systems. To add to the complexity of this zone, administrative boundaries use arbitrary lines that dissect the zone, often leading to fragmented management. This sectored approach focuses on specific activities such as land use and fisheries, often leading to adverse effects in another sector.

THE IMPORTANCE OF THE COASTAL ZONE AND THE NEED FOR MANAGEMENT

The dynamic processes that occur within the coastal zones produce diverse and productive ecosystems which have been of great importance historically for human populations. Coastal margins equate to only 8% of the worlds surface area but provide 25% of global productivity. Stress on this environment comes with approximately 70% of the world's population being within a day's walk of the coast. Two-thirds of the world's cities occur on the coast.

Valuable resources such as fish and minerals are considered to be common property and are in high demand for coastal dwellers for subsistence use, recreation and economic development. Through the perception of common property, these resources have been subjected to intensive and specific exploitation. For example; 90% of the world's fish harvest comes from within national exclusive economic zones, most of which are within the sight of shore. This type of practice has led to a problem that has cumulative effects. The addition of other activities adds to the strain placed on this environment. As a whole, human activity in the coastal zone generally degrades the systems by taking unsustainable quantities of resources. The effects are further exacerbated with the input of pollutant wastes. This provides the need for management. Due to the complex nature of human activity in this zone a holistic approach is required to obtain a sustainable outcome.

THE CONCEPT OF SUSTAINABILITY

The concept behind the idea of ICZM is sustainability. For ICZM to succeed, it must be sustainable. Sustainability entails a continuous process of decision making, so there is never an end-state just a readjustment of the equilibrium between development and the protection of the environment. The concept of Sustainability or sustainable development came to fruition in the 1987 report of the World Commission on Environment and Development, Our Common Future. It stated sustainable development is "to meet the needs of the present without compromising the ability of future generations to meet their own needs".

Highlighted are three main standpoints which summarise the idea of Sustainable development, they are:

- Economic development to improve the quality of life of people
- Environmentally appropriate development
- Equitable development

To simplify these points, sustainability should acknowledge the right of humans to live a life that is healthy and productive. It should allow for equal distribution of benefits to all people and in doing so protect the environment through appropriate use. Sustainability is by no means a set of prescriptive actions, more accurately it is a way of thinking. Adapting this way of thinking paves the way for a longer-term view with a more holistic approach, something successful ICZM can achieve.

EXPECTATIONS OF ICZM

As previously stated, for ICZM to be successful it must adhere to the principles that define sustainability and act upon them in ways that are integrated. An optimal balance between environmental protection and the development of economic and social sectors is paramount. As part of the holistic approach ICZM applies, many aspects within a coastal zone are expected to be considered and accounted for. These include but are not limited to: the spatial, functional, legal, policy, knowledge, and participation dimensions. Below are four identified goals of ICZM:

- Maintaining the functional integrity of the coastal resource systems;
- Reducing resource-use conflicts;
- Maintaining the health of the environment;
- Facilitating the progress of multisectoral development

Failure to include these aspects and goals would lead to a form of unsustainable management, undermining the paradigms explicit to ICZM.

DEFINING INTEGRATION

The term 'integration' can be adopted for many different purposes, it is therefore quite important to define the term in the context of the management of the coastal zone to appreciate the intentions of ICZM. Integration within ICZM occurs in and between many different levels, 5 types of integration that occur within ICZM, are explained below; Integration among sectors: Within the coastal environment there are many sectors that operate. These human activities are largely economic activities such as tourism, fisheries, and port companies. A sense of co-operation between sectors is the main requirement for sector integration within ICZM. This comes from the realisation of a common goal focused around sustainability and the appreciation of one another within the area.

Integration between land and water elements of the coastal zone: This is the realisation of the physical environment being a whole. The coastal environment is a dynamic relationship between many processes all of which are interdependent. The link must be made between imposing a change on one system or feature and its inevitable 'flow on' effects. Integration among levels of government: Between levels of governance, consistency and co-operation is needed throughout planning and policy making. ICZM is most effective where initiatives have common purpose at local, regional, and national levels. Common goals and actions increase efficiency and mitigate confusion.

Integration between nations: This sees ICZM as an important tool on a global scale. If goals and beliefs are common on a supranational scale, large scale problems could be mitigated or avoided.

Integration among disciplines: Throughout ICZM, knowledge should be accepted from all disciplines. All means of scientific, cultural, traditional, political and local expertise need to be accounted for. By including all these elements a truly holistic approach towards management can be achieved. The term integration in a coastal management context has many horizontal and vertical aspects, which reflects the complexity of the task and it proves a challenge to implement.

ICZM FRAMEWORK

Management must embrace a holistic viewpoint of the functions that makeup the complex and dynamic nature of interactions in the coastal environment. Management framework must be applied to a defined geographical limit (often complicated) and should operate with a high level of integration. Due to the diverse nature of the world's coastline and coastal environments, it is not possible to create a framework that is 'one-size-fits-all.' Different activities, interests and issues also complicate matters. So management will always be unique to countries, regions and ultimately on a local scale. A common thought process and decision making framework however, can be fairly uniform as a part of ICZM around the world. To achieve the principles set out in sustainable types of management a step by step process can be adhered to. Firstly, issues and problems need to be identified and assessments of these need to be quantified. This first step will include integration between government, sectoral entities and local residents. The assessments also have to be broad in their application. Once the issues and problems have been identified and weighted, an effective management plan can be made. The plan will be specific to the area in question. Thirdly, the adoption of the plan can be carried out.

They can be legally binding statutory plans, strategies or objectives which are generally quite powerful or they can be non-statutory processes and can act as a guide for future development. This duality is largely beneficial as the future can be taken into account, but still provide for a firm stance based in the present. The fourth step is implementation, this active phase includes; law enforcement, education, development etc. The implementation activities will be of course, be as unique as their environments and can take many forms. The last phase is evaluation of the whole process. The principles of sustainability mean that there is no 'end state.' ICZM is an ongoing process which should constantly readjust the equilibrium between economic development and the protection of the environment. Feedback is a crucial part of the process and allows for continued effectiveness even when a situation may change.

ICZM IN THE MEDITERRANEAN

At the Conference of the Plenipotentiaries on the ICZM Protocol that took place on 20-21 January 2008 in Madrid, the ICZM Protocol was signed. Under the presidency of the Minister of Environment of Spain, H.E. Ms. Cristina Narbona Ruiz, fourteen Contracting Parties of the Barcelona Convention signed the Protocol. These are the following: Algeria, Croatia, France, Greece, Israel, Italy, Malta, Monaco, Montenegro, Morocco, Slovenia, Spain, Syria and Tunisia. All other Parties announced to do so in the very near future. This is the 7th Protocol in the framework of the Barcelona Convention, and the decision to approve the draft text and recommendation to the Conference of the Plenipotentiaries to sign it was taken at the 15th Ordinary Meeting of the Contracting Parties during their meeting in Almeria, on 15-18 January 2008. All the parties are convinced that this Protocol is a crucial milestone in the history of MAP(UNEP/MAP). It will allow the countries to better manage their coastal zones, as well as to deal with the emerging coastal environmental challenges, such as the climate change.

The ICZM Protocol is a unique legal instrument in the entire international community and the Mediterranean countries are proud of this fact. They are willing to share these experiences with other coastal countries of the world. The signing of the Protocol came after six years of dedicated work of all the Parties.

In September 2012, Croatia and Morocco ratified the Protocol, which brings the number of ratifications to 9 (Slovenia, Montenegro, Albania, Spain, France, European Union, Syria, Croatia, Morocco). A road-map for the implementation of the ICZM Process, prepared by the Priority Action Programme (PAP/RAC) is available on the Coastal Wiki platform of the PEGASO and ENCORA projects: ICZM Process.

CONSTRAINTS OF ICZM

Major constraints of ICZM are mostly institutional, rather than technological. The 'top-down' approach of administrative decision making sees problematisation as a tool promoting ICZM through the idea of sustainability. Community-based 'bottom-up' approaches can perceive problems and issues that are specific to a local area. The benefit of this is that the problems are real and acknowledged rather than searched for to fit an imposed strategy or policy.

Public consultation and involvement is very important for current 'top-down' approaches, as it can incorporate this 'bottom-up' idea into the policies made. Prescriptive 'top-down' methods have not able to effectively address problems of resource utilization in poor coastal communities as perceptions of the coastal zone differ with regard to developed and developing countries. This leads on to another constraint to ICZM, the idea of common property. The coastal environment has huge historical and cultural connections with human activity. Its wealth of resources have provided for millennia, with regard to ICZM how does management become legally binding if the dominant perception of the coast is of a common area available to all? And should it? Enforcing restrictions or change to activities within the coastal zone can be difficult as these resources are often very important to people's livelihoods. The idea of the coast being common property fouls 'top-down' approaches. The idea of common property itself is not all that clean, This perception can lead to cumulative exploitation of resources - the very problem this management seeks to extinguish.

ICZM: THE NEW ZEALAND CASE STUDY

New Zealand is quite unique as it uses sustainable management within legislation, with a high level of importance placed on to the coastal environment. The Resource Management Act (RMA) (1991) promoted sustainable development and mandated the preparation of a New Zealand Coastal Policy Statement (NZCPS), a national framework for coastal planning. It is the only national policy statement that was mandatory. All subsequent planning must not be inconsistent with the NZCPS, making it a very important document.

Regional authorities are required to produce Regional coastal policy plans under the RMA (1991) but strangely enough, they only need to include the marine environment seaward of the mean high water mark. But many regional councils have chosen to integrate the 'dry' landward area within their plans, breaking down the artificial barriers. This attempt at ICZM is still in its early days running into many legislative hurdles and is yet to achieve a fully ecosystems-based approach. But as part of ICZM, evaluation and adoption of changes is important and ongoing changes to the NZCPS in the form of reviews is currently happening. This will provide an excellent stepping stone for future initiatives and the development of a fully integrated form of coastal management.

ICZM: THE IRAN CASE STUDY

Preparation of comprehensive management plans for optimum utilization of existent sources and potentials in all developed and developing countries is one of the appropriate approaches for constant and permanent utilization of natural, human and financial sources. The versatility of natural sources in coastal areas has made private and governmental users and investors to participate in this section to gain the utmost profits. Therefore, the necessity of preparation and implementation of management plans for perpetual utilization of existent sources in coastal areas has become inevitable. Iran, possessing some 6000 km of coastline in north and south, owns abundant economic capacities in coastal zones and regarding the versatility of nature and coast operators and management of coastal activities and operations, necessity of attention to Integrated Coastal Zone Management becomes more significant. Such necessity has gained its legal support through ratification of arrangements no. 40 from transportation chapter of third and article no. 63 of fourth economic, social and cultural economic schedule and its executive regulations. The General Director of coasts and ports engineering of Ports and Maritime Organization was detailed to take the studies of ICZM into consideration. The first phase of these studies began in spring 2003 and was fulfilled in autumn 2006. The outcome of this phase was compilation of following reports accomplished by several national and international skilled consultants: 1- Project Methodology 2- Scrutinized scope of services related to studies 3- Investigation of studies' needs and project preparation and performance 4- Study, definition and determination of Iranian coastal zones boundaries 6- Investigation of International concepts, methods and experiences about Integrated Coastal Zone Management 7- Study and investigation of different features of Integrated Coastal Zone Management in Iran 8- Preparation and designation of geographic database 9- Purchasing and preparing basic data The second phase of studies started up in autumn 2005 and since then this phase has been fully accomplished and presented, In which six competent Iranian consultants with some cooperation of international consultants are responsible for preparing the eleven results of second part of the studies.

THE EUROPEAN UNION

The European Parliament and the European Council "adopted in 2002 a Recommendation on Integrated Coastal Zone Management which defines the principles sound coastal planning and management. These include the need to base planning on sound and shared knowledge, the need to take a long-term and cross-sector perspective, to pro-actively involve stakeholders and the need to take into account both the terrestrial and the marine components of the coastal zone".

CONCLUSION

The Integrated Coastal Zone Management (ICZM) appears to be a key element for the sustainable development of these zones. However this recent notion may not be adapted to all cases. The natural disasters Sumatra earthquake and the Indian Ocean tsunami have made a lot of impact on the coastal environment and also the stakeholder's perception on mitigation and management of coastal hazards. Successful implementation is still a major challenge to the idea of ICZM .

LONGSHORE DRIFT

Longshore drift consists of the transportation of sediments (clay, silt, sand and shingle) along a coast at an angle to the shoreline, which is dependent on prevailing wind direction, swash and backwash. This process occurs in the littoral zone, and in or close to the surf zone. The process is also known as longshore transport or littoral drift. Longshore drift is influenced by numerous aspects of the coastal system, with processes that occur within the surf zone largely influencing the deposition and erosion of sediments. Longshore currents can generate oblique breaking waves which result in longshore transport.

Longshore drift affects numerous sediment sizes as it works in slightly different ways depending on the sediment (e.g. the difference in long shore drift of sediments from a sandy beach to that of sediments from a shingle beach). Sand is largely affected by the oscillatory force of breaking waves, the motion of sediment due to the impact of breaking waves and bed shear from long shore current. Whereas because shingle beaches are much steeper than sandy ones, plunging breakers are more likely to form, causing the majority of long shore transport to occur in the swash zone, due to a lack of surf zone.

LONGSHORE DRIFT FORMULAS

There are numerous calculations that take into consideration the factors that produce longshore drift. These formulations are:

1. Bijker formula (1967,1971)
2. The Engelund and Hansen formula (1967)
3. The Ackers and White formula (1973)
4. The Bailard and Inman formula (1981)
5. The Van Rijn formula (1984)
6. The Watanabe formula (1992)

These formulas all provide a different view into the processes that generate longshore drift. The most common factors taken into consideration in these formulas are:

- Suspended and bed load transport
- Waves e.g. breaking and non-breaking
- The shear exerted by waves or the flow associated with waves.

FEATURES OF SHORELINE CHANGE

Longshore drift plays a large role in the evolution of a shoreline, as if there is a slight change of sediment supply, wind direction, or any other coastal influence longshore drift can change dramatically, impacting on the formation and evolution of a beach system or profile. These changes do not occur due to one factor within the coastal system, in fact there are numerous alterations that can occur within the coastal system that may affect the distribution and impact of longshore drift. Some of these are:

1. Geological changes, e.g. erosion, backshore changes and emergence of headlands.

2. Change in hydrodynamic forces, e.g. change in wave diffraction in headland and offshore bank environments.
3. Change to hydrodynamic influences, e.g. the influence of new tidal inlets and deltas on drift.
4. Alterations of the sediment budget, e.g. switch of shorelines from drift to swash alignment, exhaustion of sediment sources.
5. The intervention of humans, e.g. cliff protection, groynes, detached breakwaters.

THE SEDIMENT BUDGET

The sediment budget takes into consideration sediment sources and sinks within a system. This sediment can come from any source with examples of sources and sinks consisting of:

- Rivers
- Lagoons
- Eroding land sources
- Artificial sources e.g. nourishment
- Artificial sinks e.g. mining/extraction
- Offshore transport
- Deposition of sediment on shore

This sediment then enters the coastal system and is transported by longshore drift. A good example of the sediment budget and longshore drift working together in the coastal system is inlet ebb-tidal shoals, which store sand that has been transported by long shore transport. As well as storing sand these systems may also transfer or by pass sand into other beach systems, therefore inlet ebb-tidal shoal systems provide a good sources and sinks for the sediment budget.

Sediment deposition (geology) throughout a shoreline profile conforms to the null point hypothesis; where gravitational and hydraulic forces determine the settling velocity of grains in a seaward fining sediment distribution. Long shore occurs in a 90 degree backwash so it would be presented as a right angle with the wave line.

SPITS

Spits are formed when longshore drift travels past a point (e.g. river mouth or re-entrant) where the dominant drift direction and shoreline do not veer in the same direction. As well as dominant drift direction, spits are affected by the strength of wave driven current, wave angle and the height of incoming waves.

Spits are landforms that have two important features, with the first feature being the region at the up-drift end or proximal end (Hart et al., 2008). The proximal end is constantly attached to land (unless breached) and may form a slight "barrier" between the sea and an estuary or lagoon. The second important spit feature is the down-drift end or distal end, which is detached from land and in some cases, may take a complex hook-shape or curve, due to the influence of varying wave directions.

As an example, the New Brighton spit in Canterbury, New Zealand, was created by longshore drift of sediment from the Waimakariri River to the north. This spit system is currently in equilibrium but undergoes phases of deposition and erosion.

BARRIERS

Barrier systems are attached to the land at both the proximal and distal end and are generally widest at the down-drift end. These barrier systems may enclose an estuary or lagoon system, like that of Lake Ellesmere enclosed by the Kaitorete Spit or hapua which form at river-coast interface such as at the mouth of the Rakaia River.

The Kaitorete Spit in Canterbury, New Zealand, is a barrier/spit system (which generally falls under the definition barrier, as both ends of the landform are attached to land, but has been named a spit) that has existed below Banks Peninsula for the last 8000 years. This system has undergone numerous changes and fluctuations due to avulsion of the Waimakariri River (which now flows to the north or Banks Peninsula), erosion and phases of open marine conditions. The system underwent further changes c.500 year BP, when longshore drift from the eastern end of the "spit" system created the barrier, which has been retained due to ongoing longshore transport.

TIDAL INLETS

The majority of tidal inlets on longshore drift shores accumulate sediment in flood and ebb shoals. Ebb-deltas may become stunted on highly exposed shores and in smaller spaces, whereas flood deltas are likely to increase in size when space is available in a bay or lagoon system. Tidal inlets can act as sinks and sources for large amounts of material, which therefore impacts on adjacent parts of the coastline.

The structuring of tidal inlets is also important for longshore drift as if an inlet is unstructured sediment may by pass the inlet and form bars at the down-drift part of the coast. Although this may also depend on the inlet size, delta morphology, sediment rate and by passing mechanism. Channel location variance and amount may also influence the impact of long shore drift on a tidal inlet as well.

For example, the Arcachon lagoon is a tidal inlet system in South west France, which provides large sources and sinks for longshore drift sediments. The impact of longshore drift sediments on this inlet system is highly influenced by the variation in the number of lagoon entrances and the location of these entrances. Any change in these factors can cause severe down-drift erosion or down-drift accretion of large swash bars.

HUMAN INFLUENCES

This section consists of long shore drift features that occur unnaturally and in some cases (e.g. groynes, detached breakwaters) have be constructed to enhance the effects of longshore drift on the coastline, but in other cases have a negative impact on long shore drift (ports and harbours).

GROYNES

Groynes are shore protection structures, placed at equal intervals along the coastline in order to stop coastal erosion and generally cross the intertidal zone. Due to this, groyne structures are usually used on shores with low net and high annual longshore drift in order to retain the sediments lost in storm surges and further down the coast.

There are numerous variations to groyne designs with the three most common designs consisting of:

1. zig-zag groynes, which dissipate the destructive flows that form in wave induced currents or in breaking waves.
2. T-head groynes, which reduce wave height through wave diffraction.
3. 'Y' head, a fish tail groyne system.

ARTIFICIAL HEADLANDS

Artificial headlands are also shore protection structures, which are created in order to provide a certain amount of protection to beaches or bays. Although the creation of headlands involves accretion of sediments on the up-drift side of the headland and moderate erosion of the down-drift end of the headland, this is undertaken in order to design a stabilised system that allows material to accumulate in beaches further along the shore.

Artificial headlands can occur due to natural accumulation or also through artificial nourishment.

DETACHED BREAKWATERS

Detached breakwaters are shore protection structures, created to build up sandy material in order to accommodate drawdown in storm conditions. In order to accommodate drawdown in storm conditions detached breakwaters have no connection to the shoreline, which lets currents and sediment pass between the breakwater and the shore. This then forms a region of reduced wave energy, which encourages the deposition of sand on the lee side of the structure. Detached breakwaters are generally used in the same way as groynes, to build up the volume of material between the coast and the breakwater structure in order to accommodate storm surges.

PORTS AND HARBOURS

The creation of ports and harbours throughout the world can seriously impact on the natural course of longshore drift. Not only do ports and harbours pose a threat to longshore drift in the short term, they also pose a threat to shoreline evolution. The major influence the creation of a port or harbour can have on longshore drift is the alteration of sedimentation patterns, which in turn may lead to accretion and/or erosion of a beach or coastal system.

As an example, the creation of a port in Timaru, New Zealand in the late 1800s led to a significant change in the longshore drift along the South Canterbury coastline. Instead of longshore drift transporting sediment north up the coast towards the Waimataitai lagoon, the creation of the port blocked the drift of these (coarse) sediments and instead caused them to accrete to the south of the port at South beach in Timaru. The accretion of this sediment to the south, therefore meant a lack of sediment being deposited on the coast near the Waimataitai lagoon (to the north of the port), which led to the loss of the barrier enclosing the lagoon in the 1930s and then shortly after, the loss of the lagoon itself. As with the Waimataitai lagoon the Washdyke Lagoon, which currently lies to the north of the Timaru port is undergoing erosion and may eventually breach causing loss of another lagoon environment.

MARINE TERRACE

A marine terrace, coastal terrace, raised beach or perched coastline is a relatively flat, horizontal or gently inclined surface of marine origin, mostly an old abrasion platform which has been lifted out of the sphere of wave activity (sometimes called "tread"). Thus it lies above or under the current sea level, depending on its time of formation. It is bounded by a steeper ascending slope on the landward side and a steeper descending slope on the seaward side (sometimes called "riser"). Due to its reasonably flat shape it is often used for anthropogenic structures like settlements and infrastructure.Morphology

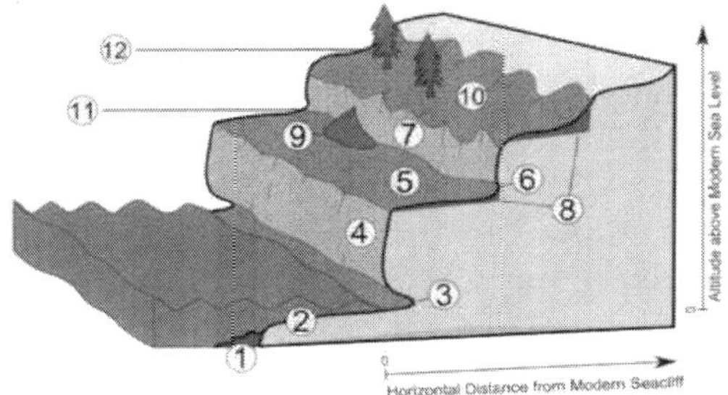

Typical sequence of erosional marine terraces. 1) low tide cliff/ramp with deposition, 2) modern shore (wave-cut/abrasion-) platform, 3) notch/inner edge, modern shoreline angle, 4) modern sea cliff, 5) old shore (wave-cut/ abrasion-) platform, 6) paleo-shoreline angle, 7) paleo-sea cliff, 8) terrace cover deposits/marine deposits, colluvium, 9) alluvial fan, 10) decayed and covered sea cliff and shore platform, 11) paleo-sea level I, 12) paleo-sea level II. - after various authors

The platform of a marine terrace usually has a gradient between 1°- 5° depending on the former tidal range with commonly a linear to concave profile. The width is very variable, reaching up to 1000 m, and seems to differ between northern and southern hemisphere. The cliff faces, delimiting the platform, can vary in steepness depending on the relative roles of marine and subaerial processes. At the intersection of the former shore (wave-cut/abrasion) platform and the rising cliff face it commonly retains a shoreline angle or inner edge (notch) which indicates the location of the shoreline at the time of maximum sea ingression and therefore a paleo sea level. Sub-horizontal platforms usually terminate in a low tide cliff and it is believed that the occurrence of these platforms depends on tidal activity. Marine terraces can extend for several tens of kilometers parallel to the coast.

Older terraces are covered by marine and/or alluvial or colluvial materials while the uppermost terrace levels usually are less well-preserved. While marine terraces in areas of relatively rapid uplift rates (> 1 mm/year) can often be correlated to individual interglacial periods or stages, those in areas of slower uplift rates may have a polycyclic origin with stages of returning sea levels after times of exposure to weathering.

Marine terraces can be covered by a wide variety of soils with complex histories and different ages. In protected areas allochtonous sandy parent materials from tsunami deposits may be found. Common soil types found on marine terraces include planosols and solonetz.

FORMATION

Causes

The formation of marine terraces is controlled by changes in environmental conditions and by tectonic activity during recent geological times. Changes in climatic conditions have led to eustatic sea-level oscillations and isostatic movements of the Earth's crust, especially with the changes between glacial and interglacial periods.

Processes of eustasy lead to glacioeustatic sea level fluctuations due to changes of the water volume in the oceans and hence to regressions and transgressions of the shoreline. At times of maximum glacial extent during the last glacial period the sea level was about 100 m lower compared to today. Eustatic sea level changes can also be caused by changes of the void volume of the oceans, either through sedimento-eustasy or tectono-eustasy.

Processes of isostasy involve the uplift of continental crusts including their shorelines. Today the process of glacial isostatic adjustment mainly applies to Pleistocene glaciated areas. In Scandinavia, for instance, the present rate of uplift reaches up to 10 mm/year. In general, eustatic marine terraces were formed during separate sea level highstands of interglacial stages and can be correlated to marine oxygene isotopic stages (MIS). Glacioisostatic marine terraces were mainly created during stillstands of the isostatic uplift. When eustasy was the main factor for the formation of marine terraces, derived sea level fluctuations can indicate former climate changes. This conclusion has to be treated with care, as isostatic adjustments and tectonic activities can be extensively overcompensated by a eustatic sea level rise. Thus, in areas of both eustatic and isostatic or tectonic influences, the course of the relative sea level curve can be complicated. Hence most of today's marine terrace sequences were formed by a combination of tectonic coastal uplift and Quaternary sea level fluctuations.

Jerky tectonic uplifts can also lead to marked terrace steps while smooth relative sea level changes may not result in obvious terraces and their formations are often not referred to as marine terraces.

Processes

Marine terraces often result from marine erosion along rocky coast lines in temperate regions due to wave attack and sediment carried in the waves. Erosion also takes place in connection with weathering and cavitation. The speed of erosion is highly dependent on the shoreline material (hardness of rock), the bathymetry and the bedrock properties and can be between only a few millimeters per year for granitic rocks and more than 10 m per year for volcanic ejecta. The retreat of the sea cliff generates a shore (wave-cut/abrasion-) platform through the process of abrasion. A relative change of the sea level leads to regressions or transgressions and eventually forms another terrace (marine-cut terrace) at a different altitude while notches in the cliff face indicate short stillstands.

It is believed that the terrace gradient increases with tidal range and decreases with rock resistance. In addition the relationship between terrace width and the strength of the rock is inverse and higher rates of uplift and subsidence as well as a higher slope of the hinterland increases the number of terraces formed during a certain time. Furthermore shore platforms are formed by denudation and marine-built terraces arise from accumulations of materials removed by shore erosion. Thus a marine terrace can be formed by both erosion and accumulation. However, there is an ongoing debate about the roles of wave erosion and weathering in the formation of shore platforms.

Reef flats or uplifted coral reefs are another kind of marine terrace found in intertropical regions. They are a result of biological activity, shoreline advance and accumulation of reef materials. While a terrace sequence can date back hundreds of thousands of years, its degradation is a rather fast process. On the one hand a deeper transgression of cliffs into the shoreline may completely destroy previous terraces; on the other hand older terraces might be decayed or covered by deposits, colluvia or alluvial fans. Erosion and backwearing of slopes caused by incisive streams play another important role in this degradation process.

MAPPING AND SURVEYING

For exact interpretations of the morphology extensive datings, surveying and mapping of marine terraces is applied. This includes stereoscopic aerial photographic interpretation (ca. 1 : 10,000 - 25,000), on-site inspections with topographic maps (ca. 1 : 10,000) and analysis of eroded and accumulated material. Moreover the exact altitude can be determined with an aneroid barometer or preferably with a levelling instrument mounted on a tripod. It should be measured with the accuracy of 1 cm and at about every 50 - 100 m, depending on the topography. In remote areas technics of photogrammetry and tacheometry can be applied.

CORRELATION AND DATING

Different methods for dating and correlation of marine terraces can be used and combined.

Correlational dating

The morphostratigraphic approach focuses especially in regions of marine regression on the altitude as the most important criterion to distinguish coast lines of different ages. Moreover individual marine terraces can be correlated based on their size and continuity. Also paleo-soils as well as glacial, fluvial, eolian and periglacial landforms and sediments may be used to find correlations between terraces. On New Zealand's North Island, for instance, tephra and loess were used to date and correlate marine terraces. At the terminus advance of former glaciers marine terraces can be correlated by their size, as their width decreases with age due to the slowly thawing glaciers along the coast line.

The lithostratigraphic approach uses typical sequences of sediment and rock strata to prove sea level fluctuations on the basis of an alternation of terrestrial and marine sediments or littoral and shallow marine sediments. Those strata show typical layers of transgressive and regressive patterns. However, an unconformity in the sediment sequence might make this analysis difficult.

The biostratigraphic approach uses remains of organisms which can indicate the age of a marine terrace. For that often mollusc shells, foraminifera or pollen are used. Especially Mollusca can show specific properties depending on their depth of sedimentation. Thus they can be used to estimate former water depths. Marine terraces are often correlated to marine oxygene isotopic stages (MIS) (e.g. Johnson, M. E.; Libbey, L. K. 1997) and can also be roughly dated using their stratigraphic position.

Direct dating

There are various methods for the direct dating of marine terraces and their related materials including C radiocarbon dating, which is the most common one. E.g. this method has been used on the North Island of New Zealand to date several marine terraces. It utilizes terrestrial biogenic materials in coastal sediments such as mollusc shells analyzing the C isotope. In some cases dating based on the Th/U ratio was applied though in case of detrital contamination or low uranium concentrations a high resolution dating was found to be difficult. In a study in southern Italy paleomagnetism was used to carry out paleomagnetic datings and luminescence dating (OSL) was used in different studies on the San Andreas Fault and on the Quaternary Eupcheon Fault in South Korea. In the last decennia, the dating of marine terraces has been enhanced since the arrival of terrestrial cosmogenic nuclides method, and particularly through the use of Be and Al cosmogenic isotopes produced in-situ. These isotopes record the duration of surface exposure to cosmic rays. and this exposure age reflects the age of abandonment of a marine terrace by the sea.

In order to calculate the eustatic sea level for each dated terrace it is assumed that the eustatic sea-level position corresponding to at least one marine terrace is known and that the uplift rate has remained essentially constant in each section.

RELEVANCE FOR OTHER RESEARCH AREAS

Marine terraces play an important role in the research on tectonics and earthquakes. They may show patterns and rates of tectonic uplift and thus may be used to estimate the tectonic activity in a certain region. In some cases the exposed secondary landforms can be correlated with known seismic events such as the 1855 Wairarapa earthquake on the Wairarapa Fault near Wellington, New Zealand which produced a 2.7 m uplift. Furthermore with the knowledge of eustatic sea level fluctuations the speed of isostatic uplift can be estimated and eventually the change of relative sea levels for certain regions can be reconstructed. Thus marine terraces also provide information for the research on climate change and trends in future sea level changes.

When analyzing the morphology of marine terraces it must be considered, that both eustasy and isostasy can have an influence on the formation process. This way can be assessed, whether there were changes in sea level or whether tectonic activities took place.

PROMINENT EXAMPLES

Marine terraces can be found on many geodynamically influenced coastlines around the world. Important sites include various coasts of New Zealand, e.g. Turakirae Head near Wellington being one of the world's best and most thoroughly studied examples. Also along the Cook Strait in New Zealand there is a well-defined sequence of uplifted marine terraces from the late Quaternary at Tongue Point. It features a well preserved lower terrace from the last interglacial, a widely eroded higher terrace from the penultimate interglacial and another still higher terrace, which is nearly completely decayed. Furthermore on New Zealand's North Island at the eastern Bay of Plenty a sequence of seven marine terraces has been studied.

Along many coasts of mainland and islands around the Pacific, marine terraces are typical coastal features. An especially prominent marine terraced coastline can be found north of Santa Cruz, near Davenport, California, where terraces probably have been raised by repeated slip earthquakes on the San Andreas Fault. Hans Jenny (pedologist) famously researched the pygmy forests of the Mendocino and Sonoma county marine terraces. The marine terrace's "ecological staircase" of Salt Point State Park is also bound by the San Andreas Fault. Along the coasts of South America marine terraces are present, where the highest ones are situated where plate margins lie above subducted oceanic ridges and the highest and most rapid rates of uplift occur. At Cape Laundi, Sumba Island, Indonesia an ancient patch reef can be found at 475 m above sea level as part of a sequence of coral reef terraces with eleven terraces being wider than 100 m.

The coral marine terraces at Huon Peninsula, New Guinea, which extend over 80 km and rise over 600 m above present sea level are currently on UNESCO's tentative list for world heritage sites under the name Houn Terraces - Stairway to the Past. Other considerable examples include marine terraces rising up to 360 m on some Philippine Islands and along the Mediterranean Coast of North Africa, especially in Tunisia, rising up to 400 m.

SUBMERSION (COASTAL MANAGEMENT)

Submersion is the sustainable cyclic portion of coastal erosion where coastal sediments move from the visible portion of a beach to the submerged nearshore region, and later return to the original visible portion of the beach. The recovery portion of the sustainable cycle of sediment behaviour is (accretion).

SUBMERSION VS EROSION

The sediment that is submerged during rough weather forms landforms including storm bars. In calmer weather waves return sediment to the visible part of the beach. Due to longshore drift some sediment can end up further along the beach from where it started. Often coastal areas have developed sustainable coastal positions where the sediment moving off beaches is sustainable submersion. Unfortunately, for many inhabited coastlines, anthropogenic interference in coastal processes has meant that erosion is often more permanent than submersion.

COMMUNITY PERCEPTION

The term erosion often is associated with undesirable impacts on the environment, whereas submersion should be celebrated as a sustainable part of healthy foreshores. Communities making decisions about coastal management need to develop understanding of the components of beach recession and be able to separate the component that is temporary sustainable submersion from the more serious irreversible anthropogenic or climate change erosion portion.

ARTIFICIAL REEF (COSTAL MANAGEMENT)

An artificial reef is a human-made underwater structure, typically built to promote marine life in areas with a generally featureless bottom, control erosion, block ship passage, or improve surfing. Many reefs are built using objects that were built for other purposes, for example by sinking oil rigs (through the Rigs-to-Reefs program), scuttling ships, or by deploying rubble or construction debris. Other artificial reefs are purpose built (e.g. the reef balls) from PVC or concrete. Shipwrecks may become artificial reefs when preserved on the sea floor. Regardless of construction method, artificial reefs generally provide hard surfaces where algae and invertebrates such as barnacles, corals, and oysters attach; the accumulation of attached marine life in turn provides intricate structure and food for assemblages of fish.

HISTORY

The construction of artificial reefs is thousands of years old. Ancient Persians blocked the mouth of the Tigris River to thwart Indian pirates by building an artificial reef, and during the First Punic War the Romans built a reef across the mouth of the Carthaginian harbor in Sicily to trap the enemy ships within and assist in driving the Carthaginians from the island.

Artificial reefs to increase fish yields or for algaculture have been used at least since 17th century Japan, when rubble and rocks were used to grow kelp, while the earliest recorded construction of artificial reef in the United States is from the 1830s when logs from huts were used off the coast of South Carolina to improve fishing.

Since at least the 1830s, American fishermen used interlaced logs to build artificial reefs. More recently, castaway junk, such as old refrigerators, shopping carts, ditched cars, out-of-service vending machines replaced the logs in ad hoc reefs. Officially sanctioned projects have incorported decommissioned subway cars, vintage battle tanks, armored personnel carriers and oil drilling rigs.

DEVELOPMENT

Artificial reefs tend to develop in more or less predictable stages. First, where an ocean current encounters a vertical structure, it can create a plankton-rich upwelling that provides a reliable feeding spot for small fish such as sardines and minnows, which draw in pelagic predators like bluefin tuna and sharks. Next come creatures seeking protection from the ocean's lethal openness-hole and crevice dwellers such as grouper, snapper, squirrelfish, eels, and triggerfish. Opportunistic predators such as jack and barracuda also appear, waiting for their prey to venture out. Over months and years the reef structure becomes encrusted with algae, tunicates, hard and soft corals, and sponges.

REEFS

Florida

Florida is the site of many artificial reefs, many created from deliberately sunken ships, including Coast Guard cutters Duane and Bibb and the U.S. Navy landing ship Spiegel Grove.

Osborne Reef

In the early 1970s, more than 2,000,000 used vehicle tires were dumped off the coast of Fort Lauderdale, Florida to form an artificial reef. Catastrophically, the tires were not properly secured to the reef structures. Ocean currents broke them loose, sending them crashing into the developing reef and its natural neighbors. As of 2009, less than 100,000 of the tires had been removed after more than 10 years of efforts.

Neptune Reef

Neptune Reef was originally conceived as an art project that would gradually decay. Burial at sea became a way of financing the project. As of 2011, about 200 "placements" had taken place. Cremated remains are mixed with cement and either encased in columns or molded into sea-star, brain-coral, 15 foot-tall castings of lions or other shapes before entering the water.

The ex-USS Oriskany

The world's largest artificial reef was created by the purposeful sinking of the aircraft carrier USS Oriskany off the coast of Pensacola, Florida, in 2006.

The ex-USNS Hoyt S. Vandenberg

The second-largest artificial reef is the USNS Hoyt S. Vandenberg, a former World War II era troop transport that served as a spacecraft tracking ship after the war. The Vandenberg was scuttled seven miles off Key West on May 27, 2009, in 140 feet of clear water. Supporters expect the ship to draw recreational divers away from natural reefs, allowing those reefs to recover from damage from overuse.

The ex-USS Spiegel Grove

The ex-Spiegel Grove is located on Dixie Shoal, 6 miles (10 km) off the Florida Keys in the Florida Keys National Marine Sanctuary. Her exact location is 25°04?00.2?N 80°18?00.7?W.

DELAWARE

Redbird Reef

In late 2000, The MTA New York City Transit decided to phase out its outdated fleet of subway cars to make room for the R142 and R142A trains. These subway cars, (Redbirds), ran on the IRT lines in the New York City Subway system for 40 years. Each car was sold, stripped, decontaminated, loaded on a barge, and sunk in the Atlantic Ocean. Some had number plates removed because of rust, and auctioned off on eBay. 1,200 subway cars were sunk. In September 2007 the MTA approved a contract worth $6 million, to send 1,600 of its retired subway cars to be used as artificial reefs. Most of these trains ran on The BMT/IND lines. The trains include the R32, R38, R40 and R42. These models are stainless steel. The MTA will replace them with the R160A and R160B trains. The plastic front ends were removed before sinking. The retired fleet included old work trains and cars damaged beyond repair.

MEXICO

Cancun Underwater Museum

Since November 2009, artist Jason Decaires Taylor has created more than 400 life size sculptures off the coast of Cancun, Mexico. The coral reefs in this region suffered heavy degradation due to repetitive hurricane abuse. This project funded by The National Marine Park and the Cancun Nautical Association was designed to emmulate coral reefs using a neutral ph clay. Taylor has constructed unique settings depicting daily activities ranging from a man watching TV to a 1970's replica of a Volkswagen Beetle. This artificial reef has relieved pressure from the nearby Manchones Reef. The design and materials implemented in this project have proved the ecological viability of artificial reefs.

AUSTRALIA

Since the late 1990s, the Australian government has been providing decommissioned warships for use as artificial reefs for recreational scuba diving. So far, the following six ships have been sunk:

- Ex-HMAS Swan at Dunsborough in Western Australia during December 1998.
- Ex-HMAS Perth at Albany in Western Australia during November 2001.
- Ex-HMAS Hobart in Yankalilla Bay in South Australia during November 2002.
- Ex-HMAS Brisbane off the Sunshine Coast in Queensland during July 2005.

- Ex-HMAS Canberra at a site west of the entrance to Port Phillip Bay in Victoria during October 2009.
- Ex-HMAS Adelaide off Terrigal on the New South Wales Central Coast during April 2011

GIBRALTAR

The Gibraltar Reef was first proposed by Eric Shaw in 1973. Initial experiments with tyres proved unsuccessful as the tyres were simply swept away by currents or buried underneath sand. In 1974, boats from local marinas and the Gibraltar Port Authority were donated. The first two were barges that were sunk in Camp Bay. In 2006, a 65 ton wooden boat, "True Joy" (also referred to as "Noah's Ark") was sunk here as well, followed by the MV New Flame, a mid sized bulk carrier, in 2007. In 2013, a dropping of more than 70 concrete blocks of one square meter with metal bars, took place. There was a discussion about, because some experts don't consider the blocks of concrete with metal bars a real reef but a system to bother Spanish fishing activities. The blocks with metal devices proved unpopular with Spanish fishermen who have suffered important economic loses and demanded the process to be stopped because it would prevent them trawling the seabed for fish. The metal bars are designed to break the fishermen nets. The dropping lead to a diplomatic conflict between the Kingdom of Spain and United Kingdom.

ARTIFICIAL SURFING REEFS

Artificial surfing reefs have been created for surfing, coastal protection, habitat enhancement and coastal research. The world's first attempt was made in El Segundo, near Los Angeles, in California. The next attempt was at Mosman Beach, Perth, Western Australia. This reef was constructed of large granite rocks placed in a pyramidal shape to form an appropriate breaking wave form that would suit surfers. An artificial reef constructed of over 400 massive, geotextile bags (each one larger than a bus) filled with sand was constructed in 2000 at Narrowneck on the Gold Coast of Queensland, Australia. This artificial reef had two objectives: stabilizing beach nourishment and improving surfing conditions.

In the United States, in particular, demanding coastal permit requirements and environmental opposition present major obstacles to building surfing reefs. As of February 2006, the only reef built in the U.S. for surfing is southern California's "Pratte's Reef", which failed to create waves. Reefs built to enhance marine habitat face less environmental opposition, in part because they are in deeper water and further offshore. A number of such man-made reefs exist near Florida and Hawaii. Artificial surfing reefs typically resemble a "submerged breakwater", and proponents suggest benefits beyond surfing conditions. Many coastlines are subject to powerful waves that crash directly onshore. An artificial reef 150-300 yards offshore might create surfing opportunities and, by dissipating wave energy, make swimming safer and reduce coastal erosion. The USS Spiegel Grove was sunk in 2002 to make an artificial reef. Europe's first artificial reef was approved in 2008. Construction began August 30, 2008 in Boscombe, Bournemouth, UK, and opened in November 2009. The multi-purpose reef reef is expected to create waves up to 30% larger and double the number of surfing days annually. Construction on this reef began in June 2008, and was completed in August 2009. Boscombe Reef was built from large sand-filled geotextile containers, totaling 13,000 cubic meters.

ELECTRO MINERAL ACCRETION (EMA)

Mineral accretion involves applying a low voltage current to a metallic structure to cause limestone to crystallize on the surface, to which coral planulae can attach and grow. The electric current also speeds post-attachment growth. EMA works like charging a battery with a positive pole, the cathode, and a negative pole, the anode. Applying electric current attracts various dissolved minerals to either the cathode or the anode. Chemical reactions take place at both poles. On the anode, bubbles of oxygen and chlorine gas form. These bubbles float to the surface and dissolve into the air. On the cathode, bubbles of hydrogen gas and a limestone precipitate appear.

The voltage is low enough that it can be generated by floating solar panels or from wave motion. A coalition of scientists named the Global Coral Reef Alliance (GCRA) is developing a technique called the Biorock Process using mineral accretion for reef restoration, mariculture, and shoreline protection.

ENVIRONMENTAL CONCERNS

According to The Ocean Conservancy, a Washington-based environmental group, the Osbourne reef may be an indication that the benefits of artificial reefs need to be re-examined. Jack Sobel, a senior scientist at the group, has said "There's little evidence that artificial reefs have a net benefit," citing concerns such as toxicity, damage to ecosystems and concentrating fish into one place (worsening overfishing).

ACCRETION (COASTAL MANAGEMENT)

Accretion is the process of coastal sediment returning to the visible portion of a beach or foreshore following a submersion event. A sustainable beach or foreshore often goes through a cycle of submersion during rough weather then accretion during calmer periods. If a coastline is not in a healthy sustainable state, then erosion can be more serious and accretion does not fully restore the original volume of the visible beach or foreshore leading to permanent beach loss.

MANAGED RETREAT

In the context of coastal erosion, managed retreat (also managed realignment) allows an area that was not previously exposed to flooding by the sea to become flooded by removing coastal protection. This process is usually in low lying estuarine areas and almost always involves flooding of land that has at some point in the past been claimed from the sea. Managed retreat is often a response to sea level rise exacerbated by local subsidence of the land surface due to isostatic rebound in the north.

COASTAL DEFENCE

In the UK the main reason for implementation of Managed Realignment is generally to improve coastal stability, essentially replacing artificial 'hard' coastal defences with natural 'soft' coastal landforms (Pethick 2002). This process can be used to protect areas of land further inland rather than that near the coast by relying on natural defences to absorb or dampen the force of waves.

HABITAT LOSS

In addition to being used as a means of coastal defence, Managed Realignment has also been used in a number of cases to mitigate for loss of intertidal habitat. Although land claim has been an important factor for salt marsh loss in the UK in the past (Allen 1992) the majority of current salt marsh loss in the UK is believed to be due to erosion (Morris et al. 2004). This erosion may involve coastal squeeze, where protective sea walls prevent the landward migration of salt marsh in response to sea level rise when sediment supply is limited (Hulme 2005; Morris et al. 2004). Salt marshes are protected under the EU Habitats Directive as well as providing habitat for a number of species protected by the Birds Directive. Following this guidance, the UK's biodiversity action plan aims to prevent net losses to the area of salt marsh present in 1992. It is therefore a legal requirement that all losses in marsh area must be compensated by replacement habitat with equivalent biological characteristics (Crooks et al. 2001). This equates to the need to restore approximately 1.4 km^2 of salt marsh habitat per year in the UK.

ADVANTAGES

There are no direct costs apart from that of removing any defences already in place and maintenance costs are very low. Sediment flow is also restored to its natural state, beaches can be naturally replenished due to erosion of the coast, providing protection and the balance of the coastline returns.

DISADVANTAGES

A certain amount of land will inevitably be lost in this process while beaches are being built up resulting in settlements, farmland and other property being destroyed. Because of this, managed retreat is often not a socially acceptable plan and may invoke the need for compensation to land-owners. Intertidal sites are often a rich archaeological resource and the loss of heritage is a factor to be weighed in managed retreat projects. There are no agreed protocols on the monitoring of MR sites (Atkinson et al. 2001) and, consequently, very few of the sites are being monitored consistently and effectively (Wolters et al. 2005c). Due to the low levels of monitoring there is little evidence on which to base future managed realignment projects. This has led to the results of Managed Realignment schemes being extremely unpredictable.

EXAMPLES

In the UK, the first managed retreat site was an area of 8,000 square metres at Northey Island in Essex flooded in 1991, followed by larger sites at Tollesbury and Orplands (1995), Freiston Shore (2001) and Abbott's Hall Farm, at Great Wigborough in the Blackwater Estuary, it is one of the largest managed retreat schemes in Europe. It covers nearly 280 hectares of land on the north side of the estuary (2002) and a number of others. The programme was started by the RSPB - The Royal Society for the Protection of Birds, of whom own Abbott's Hall Farm, they made five breaches in the original old sea wall to allow the held-back sea to flood through to create salt marshland. The marshland over time reverted to its original state before the time of being cultivated, it has become a great site for birds as for the perfect conditions and is a hot-spot breeding ground too.

CURRENT PROGRESS

At present approximately 4 km^2 of salt marsh have been restored by MR in the UK (Mossman et al. In prep). One of the major reasons cited for the slow pace of current salt marsh restoration in the UK (Morris et al. 2004) is the uncertainty associated with the practice (Foresight).

INTERTIDAL ZONE

The intertidal zone, also known as the foreshore and seashore and sometimes referred to as the littoral zone, is the area that is above water at low tide and under water at high tide (in other words, the area between tide marks). This area can include many different types of habitats, with many types of animals, such as starfish, sea urchins, and numerous species of coral. The well-known area also includes steep rocky cliffs, sandy beaches, or wetlands (e.g., vast mudflats). The area can be a narrow strip, as in Pacific islands that have only a narrow tidal range, or can include many meters of shoreline where shallow beach slopes interact with high tidal excursion. Organisms in the intertidal zone are adapted to an environment of harsh extremes. Water is available regularly with the tides but varies from fresh with rain to highly saline and dry salt with drying between tidal inundations. The action of waves can dislodge residents in the littoral zone. With the intertidal zone's high exposure to the sun, the temperature range can be anything from very hot with full sun to near freezing in colder climates. Some microclimates in the littoral zone are ameliorated by local features and larger plants such as mangroves. Adaptation in the littoral zone allows the use of nutrients supplied in high volume on a regular basis from the sea which is actively moved to the zone by tides. Edges of habitats, in this case land and sea, are themselves often significant ecologies, and the littoral zone is a prime example.

A typical rocky shore can be divided into a spray zone or splash zone (also known as the supratidal zone), which is above the spring high-tide line and is covered by water only during storms, and an intertidal zone, which lies between the high and low tidal extremes. Along most shores, the intertidal zone can be clearly separated into the following subzones: high tide zone, middle tide zone, and low tide zone. The intertidal zone is one of a number of marine biomes or habitats, including estuaries, neritic, surface and deep zones.

ZONATION

Marine biologists divide the intertidal region into three zones (low, middle, and high), based on the overall average exposure of the zone. The low intertidal zone, which borders on the shallow subtidal zone, is only exposed to air at the lowest of low tides and is primarily marine in character. The mid intertidal zone is regularly exposed and submerged by average tides. The high intertidal zone is only covered by the highest of the high tides, and spends much of its time as terrestrial habitat.

The high intertidal zone borders on the splash zone (the region above the highest still-tide level, but which receives wave splash). On shores exposed to heavy wave action, the intertidal zone will be influenced by waves, as the spray from breaking waves will extend the intertidal zone. Depending on the substratum and topography of the shore, additional features may be noticed. On rocky shores, tide pools form in depressions that fill with water as the tide rises. Under certain conditions, such as those at Morecambe Bay, quicksand may form.

LOW TIDE ZONE (LOWER LITTORAL)

This subregion is mostly submerged - it is only exposed at the point of low tide and for a longer period of time during extremely low tides. This area is teeming with life; the most notable difference with this subregion to the other three is that there is much more marine vegetation, especially seaweeds. There is also a great biodiversity. Organisms in this zone generally are not well adapted to periods of dryness and temperature extremes. Some of the organisms in this area are abalone, sea anemones, brown seaweed, chitons, crabs, green algae, hydroids, isopods, limpets, mussels, nudibranchs, sculpin, sea cucumber, sea lettuce, sea palms, sea stars, sea urchins, shrimp, snails, sponges, surf grass, tube worms, and whelks. Creatures in this area can grow to larger sizes because there is more available energy in the localized ecosystem. Also, marine vegetation can grow to much greater sizes than in the other three intertidal subregions due to the better water coverage. The water is shallow enough to allow plenty of light to reach the vegetation to allow substantial photosynthetic activity, and the salinity is at almost normal levels. This area is also protected from large predators such as fish because of the wave action and the relatively shallow water.

MARINE FUEL MANAGEMENT

Marine fuel management (MFM) is a multi-level approach to measuring, monitoring, and reporting fuel usage on a boat or ship, with the goals of reducing fuel usage, increasing operational efficiency, and improving fleet management oversight. MFM has grown in importance due to the rising costs of marine fuel and increased governmental pressures to reduce the pollution generated by the world's fleet.

Effective MFM requires that you know:

- How much fuel is used
- How the fuel was used
- What things impact fuel usage
- And by how much

Manual methods of measuring fuel usage, i.e. fuel tank dipping or sounding, typically do not tell how much fuel was used:

- Traveling versus idling while in port or on station
- By a specific engine (port versus starboard, for example)
- Performing one job versus another
- By crew A versus crew B on similar voyages

Without a clear understanding of how fuel is being used, there is no operational baseline from which to compare any kind of fuel conservation tool or activity. Without a baseline, there is no way to determine if conservation strategies are actually working.

MFM allows a fleet owner to track actual fuel consumption and relate fuel consumption to the work performed by the vessel. It supports the analysis of the effectiveness of operating strategies and helps develop a clearer understanding of how well a vessel uses its fuel.

MFM MAIN FUNCTIONAL AREAS

- Operational Performance
- Engineering and Maintenance Management
- Management Oversight

OPERATIONAL PERFORMANCE

Operational performance includes those functional areas that impact the actual performance of a vessel or fleet. It includes fuel monitoring, inventory control, accounting, and engine throttle management.

FUEL MONITORING

Many marine vessels do not provide a way for captain and crew to measure and monitor fuel usage while underway. An optimum system onboard would include the ability to instantaneously monitor fuel burn rates from the wheelhouse. Individual engine and generator burn rates would be included, as well as fuel tank levels. This proactive monitoring would allow the crew to make decisions that positively impact fuel burn rates and efficiency.

INVENTORY CONTROL

Fuel tanks need to have sensors installed that continuously monitor levels as fuel is taken onboard and burned by engines and generators. Periodically measuring tank levels using traditional manual methods is not accurate enough or timely, given the volumes of fuel that a marine engine can consume. Meters should be installed on transfer lines where fuel is taken onboard or off-loaded.

ACCOUNTING

In some parts of the world, fuel theft is an ongoing concern. Consequently, the accurate measurement of fuel taken on board coupled with the fuel actually consumed by engines and generators, is an important part of MFM. Meters should be installed in all fuel transfer lines so accurate fueling data can be captured. This data can then be compared with burn rates to determine whether fuel is being transferred off the vessel secretively.

Beyond fuel theft, many governmental jurisdictions require that all fuel spill incidents be recorded and reported to the local authorities. For example, the Marine Department of the Government of Hong Kong has specific guidelines for responding to accidental marine fuel spills which reflect international requirements as promulgated by MARPOL, the International Convention for the Prevention of Pollution From Ships. Additionally, accounting for fuel usage at various points along a voyage provides the ability to tie fuel burn and its associated costs to shipping or container rates. For example, understanding how a vessel burns fuel on certain parts of a voyage, allows the more accurate bidding of container rates so profit margins stay healthy.

Consequently, varying shipping rates based on documented fuel usage rates can allow a shipper to bid more aggressively. A modern marine fuel management system would help in monitoring fuel usage, fuel transfers, bunkering events and could be configured to sound an audible alarm when refilling fuel tanks might lead to a spill.

THROTTLE MANAGEMENT

Vessel operators have the most control over fuel usage by the way they use the engine(s) throttle. Wind, current, hull condition, load, and propulsion system health can all impact fuel burn both positively or negatively. Some operators choose to lower engine speed, and hence vessel speed, in an attempt to save fuel.

However, engine RPM and vessel speed alone are not indicative of total fuel consumption, so arbitrarily lowering engine speed does not guarantee fuel savings. One must do the workflow calculations on how the propulsion system is operating under existing changing conditions and then tie that to fuel consumption. Simply lowering engine RPM does not guarantee an optimum vessel speed setting based on conditions. Some modern fuel management systems are designed to perform these calculations while underway and make recommendations to the vessel master.

ENGINEERING AND MAINTENANCE MANAGEMENT

As with any capital asset, manufacturers typically include the standard maintenance practices and procedures needed to keep the asset functioning properly and within design specifications. In many cases, scheduled maintenance routines are based on laboratory or design parameters and do not necessarily represent the optimum.

MFM supports proper maintenance on marine engines and generators by using the actual fuel burned or hours operated as the basis for performing maintenance routines. This condition-based maintenance program more accurately reflects the operating environment of the engine, but more importantly, reduces or eliminates unnecessary maintenance work.

MANAGEMENT OVERSIGHT

Management functions within MFM include:

- Vessel performance analysis and overall fleet performance
- Crew analysis with emphasis on applying lessons learned as best practice across the fleet
- Key Performance Indicator (KPI) gathering across the fleet to include fuel burned per mile or per ton; throttle settings at various points on a voyage; engine RPMs and exhaust gas analysis; and vessel performance against hull conditions.
- Fuel management from purchasing to transfers to usage
- Chartered vessel fuel performance and adherence to contractual obligations

13

Deep Sea Mining

Deep sea mining is a relatively new mineral retrieval process that takes place on the ocean floor. Ocean mining sites are usually around large areas of polymetallic nodules or active and extinct hydrothermal vents at about 1,400 - 3,700 m below the ocean's surface.

The vents create sulfide deposits, which contain valuable metals such as silver, gold, copper, manganese, cobalt, and zinc. The deposits are mined using either hydraulic pumps or bucket systems that take ore to the surface to be processed. As with all mining operations, deep sea mining raises questions about potential environmental impact on surrounding areas. Environmental advocacy groups such as Greenpeace have argued that seabed mining should not be permitted in most of the world's oceans because of the potential for damage to deepsea ecosystems.

BRIEF HISTORY

In the mid 1960s the prospect of deep-sea mining was brought up by the publication of J. L. Mero's Mineral Resources of the Sea. The book claimed that nearly limitless supplies of cobalt, nickel and other metals could be found throughout the planet's oceans. Mero stated that these metals occurred in deposits of manganese nodules, which appear as lumps of compressed sediment on the sea floor at depths of about 5,000 m. Some nations including France, Germany and the United States sent out research vessels in search of nodule deposits. Initial estimates of deep sea mining viability turned out to be much exaggerated. This overestimate, coupled with depressed metal prices, led to the near abandonment of nodule mining by 1982. From the 1960s to 1984 an estimated US $650 million had been spent on the venture, with little to no return.

Over the past decade a new phase of deep-sea mining has begun. Rising demand for precious metals in Japan, China, Korea and India has pushed these countries in search of new sources. Interest has recently shifted toward hydrothermal vents as the source of metals instead of scattered nodules. The trend of transition towards an electricity based information and transportation infrastructure currently seen in western societies further pushes demands for precious metals. The current revived interest in phosphorus nodule mining at the seafloor stems from phosphor-based artificial fertilizers being of significant importance for world food production. Growing world population pushes the need for artificial fertilizers or greater incorporation of organic systems within agricultural infrastructure.

Currently, the best potential deep sea site, the Solwara 1 Project, has been found in the waters off Papua New Guinea, a high grade copper-gold resource and the world's first Seafloor Massive Sulphide (SMS) resource. The Solwara 1 Project is located at 1600 metres water depth in the Bismarck Sea, New Ireland Province. Using ROV (remotely operated underwater vehicles) technology, Nautilus Minerals Inc. is first company of its kind to announce plans to begin full-scale undersea excavation of mineral deposits. However a dispute with the government of Papua-New Guinea has delayed production from its expected start in early 2013.

LAWS AND REGULATIONS

The international law based regulations on deep sea mining are contained in the United Nations Conventions on the Law of the Sea from 1973 to 1982, which came into force in 1994. The convention set up the International Seabed Authority (ISA), which regulates nations' deep sea mining ventures outside each nations' Exclusive Economic Zone (a 200-nautical-mile (370 km) area surrounding coastal nations). The ISA requires nations interested in mining to explore two equal mining sites and turn one over to the ISA, along with a transfer of mining technology over a 10 to 20 year period.

This seemed reasonable at the time because it was widely believed that nodule mining would be extremely profitable. However, these strict requirements led some industrialized countries to refuse to sign the initial treaty in 1982.

RESOURCES MINED

The deep sea contains many different resources available for extraction, including silver, gold, copper, manganese, cobalt, and zinc. These raw materials are found in various forms on the sea floor, usually in higher concentrations than terrestrial mines.

Minerals and related depths

Type of mineral deposit	Average Depth	Resources found
Polymetallic nodules	4,000 - 6,000 m	Nickel, copper, cobalt, and manganese
Manganese Crusts	800 - 2,400 m	Mainly cobalt, some vanadium, molybdenum and platinum
Sulfide deposits	1,400 - 3,700 m	Copper, lead and zinc some gold and silver

Diamonds are also mined from the seabed by De Beers and others. Nautilus Minerals Inc and Neptune Minerals are planning to mine the offshore waters of Papua New Guinea and New Zealand.

EXTRACTION METHODS

Recent technological advancements have given rise to the use remotely operated vehicles (ROVs) to collect mineral samples from prospective mine sites. Using drills and other cutting tools, the ROVs obtain samples to be analyzed for precious materials. Once a site has been located, a mining ship or station is set up to mine the area.

There are two predominant forms of mineral extraction being considered for full scale operations: continuous-line bucket system (CLB) and the hydraulic suction system. The CLB system is the preferred method of nodule collection. It operates much like a conveyor-belt, running from the sea floor to the surface of the ocean where a ship or mining platform extracts the desired minerals, and returns the tailings to the ocean. Hydraulic suction mining lowers a pipe to the seafloor which transfers nodules up to the mining ship. Another pipe from the ship to the seafloor returns the tailings to the area of the mining site.

In recent years, the most promising mining areas have been the Central and Eastern Manus Basin around Papua New Guinea and the crater of Conical Seamount to the east. These locations have shown promising amounts of gold in the area's sulfide deposits (an average of 26 parts per million). The relatively shallow water depth of 1050 m, along with the close proximity of a gold processing plant makes for an excellent mining site.

ENVIRONMENTAL IMPACTS

Because deep sea mining is a relatively new field, the complete consequences of full scale mining operations are unknown. However, some researchers have said they believe that removal of parts of the sea floor will result in disturbances to the benthic layer, increased toxicity of the water column and sediment plumes from tailings. Removing parts of the sea floor could disturb the habitat of benthic organisms, with unknown long-term effects. Aside from the direct impact of mining the area, some researchers and environmental activists have raised concerns about leakage, spills and corrosion that could alter the mining area's chemical makeup.

Among the impacts of deep sea mining, sediment plumes could have the greatest impact. Plumes are caused when the tailings from mining (usually fine particles) are dumped back into the ocean, creating a cloud of particles floating in the water. Two types of plumes occur: near bottom plumes and surface plumes. Near bottom plumes occur when the tailings are pumped back down to the mining site. The floating particles increase the turbidity, or cloudiness, of the water, clogging filter-feeding apparatuses used by benthic organisms. Surface plumes cause a more serious problem. Depending on the size of the particles and water currents the plumes could spread over vast areas. The plumes could impact zooplankton and light penetration, in turn affecting the food web of the area.

CURRENT EVENTS AND NEWS

The Nautilus Minerals Corporation is currently in talks with the government of Papua New Guinea, attempting to setup and start a deep sea mining project.

SUBSEA (TECHNOLOGY)

Subsea is a general term frequently used to refer to equipment, technology, and methods employed in marine biology, undersea geology, offshore oil and gas developments, underwater mining, and offshore wind power industries.

OIL AND GAS

Oil and gas fields reside beneath many inland waters and offshore areas around the world, and in the oil and gas industry the term subsea relates to the exploration, drilling and development of oil and gas fields in underwater locations. Under water oil field facilities are generically referred to using a subsea prefix, such as subsea well, subsea field, subsea project, and subsea development.

Subsea oil field developments are usually split into Shallow water and Deepwater categories to distinguish between the different facilities and approaches that are needed. The term shallow water or shelf is used for shallow water depths where bottom-founded facilities like jackup drilling rigs and fixed offshore structures can be used, and where saturation diving is feasible.

Deepwater is a term often used to refer to offshore projects located in water depths greater than around 600 feet, where floating drilling vessels and floating oil platforms are used, and remotely operated underwater vehicles are required as manned diving is not practical. Subsea completions can be traced back to 1943 with the Lake Erie completion at a 35-ft water depth. The well had a land-type Christmas tree that required diver intervention for installation, maintenance, and flow line connections.

Shell completed its first subsea well in the Gulf of Mexico in 1961. The first known subsea ultra-high pressure waterjet system capable of operating below 5,000 ft was developed in 2010 by Jet Edge and Chukar Waterjet. It was used to blast away hydrates that were clogging a containment system at the Gulf oil spill site. Chukar Waterjet has since developed a deepwater subsea waterjet system capable of operating at depths of up to 3000 meters (10,000 feet).Effective at cutting steel up to 250 mm thick or waterjet blasting at pressures up to 3800 bar, the system can be used to blast away coatings and marine growth to inspect welds, or as a cutting tool in emergency response and salvage operations. It also can be used for hydrate remediation.

SYSTEMS

Subsea production systems can range in complexity from a single satellite well with a flowline linked to a fixed platform, FPSO or an onshore installation, to several wells on a template or clustered around a manifold, and transferring to a fixed or floating facility, or directly to an onshore installation. Subsea production systems can be used to develop reservoirs, or parts of reservoirs, which require drilling of the wells from more than one location.

Deep water conditions, or even ultradeep water conditions, can also inherently dictate development of a field by means of a subsea production system, since traditional surface facilities such as on a steel-piled jacket, might be either technically unfeasible or uneconomical due to the water depth. The development of subsea oil and gas fields requires specialized equipment. The equipment must be reliable enough to safe guard the environment, and make the exploitation of the subsea hydrocarbons economically feasible.

The deployment of such equipment requires specialized and expensive vessels, which need to be equipped with diving equipment for relatively shallow equipment work (i.e. a few hundred feet water depth maximum), and robotic equipment for deeper water depths. Any requirement to repair or intervene with installed subsea equipment is thus normally very expensive. This type of expense can result in economic failure of the subsea development.

Subsea technology in offshore oil and gas production is a highly specialized field of application with particular demands on engineering and simulation. Most of the new oil fields are located in deepwater and are generally referred to as deepwater systems. Development of these fields sets strict requirements for verification of the various systems' functions and their compliance with current requirements and specifications. This is because of the high costs and time involved in changing a pre-existing system due to the specialized vessels with advanced onboard equipment.

A full scale test (System Integration Test - SIT) does not provide satisfactory verification of deepwater systems because the test, for practical reasons, cannot be performed under conditions identical to those under which the system will later operate. The oil industry has therefore adopted modern data technology as a tool for virtual testing of deepwater systems that enables detection of costly faults at an early phase of the project.

By using modern simulation tools models of deepwater systems can be set up and used to verify the system's functions, and dynamic properties, against various requirements specifications. This includes the model-based development of innovative high-tech plants and system solutions for the exploitation and production of energy resources in an environmentally friendly way as well as the analysis and evaluation of the dynamic behavior of components and systems used for the production and distribution of oil and gas. Another part is the real-time virtual test of systems for subsea production, subsea drilling, supply above sea level, seismography, subsea construction equipment and subsea process measurement and control equipment.

OFFSHORE WIND POWER

The power transmission infrastructure for offshore wind power utilizes a variety of subsea technologies for the installation and maintenance of submarine power transmission cables and other electrical energy equipment. In addition, the monopile foundations of fixed-bottom wind turbines and the anchoring and cable structures of floating wind turbines are regularly inspected with a variety of shipborne subsea technology.

UNDERWATER MINING

Recent technological advancements have given rise to the use of remotely operated vehicles (ROVs) to collect mineral samples from prospective mine sites. Using drills and other cutting tools, the ROVs obtain samples to be analyzed for desired minerals. Once a site has been located, a mining ship or station is set up to mine the area.

REMOTELY OPERATED VEHICLES

Remotely Operated Vehicles (ROVs) are robotic pieces of equipment operated from afar to perform tasks on the sea floor. ROVs are available in a wide variety of function capabilities and complexities from simple "eyeball" camera devices, to multi-appendage machines that require multiple operators to operate or "fly" the equipment.

ORGANIZATIONS

A number of professional societies and trade bodies are involved with the subsea industry around the world. Such groups include

- Society for Underwater Technology
- Subsea Project Awards
- Subsea UK
- Subsea Valley The largest subsea technology business cluster.
- Society of Petroleum Engineers
- American Petroleum Institute (API)
- American Society of Mechanical Engineers (ASME)
- National Association of Corrosion Engineers (NACE).

Government agencies administer regulations in their territorial waters around the world. Examples of such government agencies are the Minerals Management Service (MMS, US), Norwegian Petroleum Directorate (NPD, Norway), and Health & Safety Executive (HSE, UK). The MMS administers the mineral resources in the US (using Code of Federal Regulations (CFR)) and provides management of all the US subsea mineral and renewable energy resources.

14

Submarine Pipeline

A submarine pipeline is a pipeline that is laid on the seabed or below it inside a trench. In some cases, the pipeline is mostly on-land but in places it crosses water expanses, such as small seas, straights and rivers. Submarine pipelines are used primarily to carry oil or gas, but transportation of water is also important.

A distinction is made between a flowline and a pipeline. The former is an intrafield pipeline, in the sense that it is used to connect subsea wellheads, manifolds and the platform within a particular development field. The latter, sometimes referred to as an export pipeline, is used to bring the resource to shore. Sizeable pipeline projects need to take into account a large number of factors, such as the offshore ecology, geohazards and environmental loading - they are often undertaken by multidisciplinary, international teams.

ROUTE SELECTION

One of the earliest and most critical tasks in a submarine pipeline planning exercise is the route selection. This selection has to consider a variety of issues, some of a political nature, but most others dealing with geohazards, physical factors along the prospective route, and other uses of the seabed in the area considered. This task begins with a fact-finding exercise, which is a standard desk study that includes a survey of geological maps, bathymetry, fishing charts, aerial and satellite photography, as well as information from navigation authorities.

PHYSICAL FACTORS

The primary physical factor to be considered in submarine pipeline construction is the state of the seabed - whether it is smooth (i.e. relatively flat) or uneven (bumpy, with high points and low points). If it is uneven, the pipeline will include free spans when it connects two high points, leaving the section in between unsupported. If an unsupported section is too long, the bending stress exerted onto it (due to its weight) may be excessive. Vibration from current-induced vortexes may also become an issue. Corrective measures for unsupported pipeline spans include seabed leveling and post-installation support, such as berm or sand infilling below the pipeline. The strength of the seabed is another significant parameter. If the soil is not strong enough, the pipeline may sink into it to an extent where inspection, maintenance procedures and prospective tie-ins become difficult to carry out. At the other extreme, a rocky seabed is expensive to trench and, at high points, abrasion and damage of the pipeline's external coating may occur. Ideally, the soil should be such as to allow the pipe to settle into it to some extent, thereby providing it with some lateral stability.

Other physical factors to be taken into account prior to building a pipeline include the following:

- Seabed mobility: Sandwaves and megaripples are features that move with time, such that a pipeline that was supported by the crest of one such feature during construction may find itself in a through later during the pipeline's operational lifespan. The evolution of these features is difficult to predict so it is preferable to avoid the areas where they are known to exist.
- Submarine landslides: They result from high sedimentation rates and occur on steeper slopes. They can be triggered by earthquakes. When the soil around the pipe is subjected to a slide, especially if the resulting displacement is at high angle to the line, the pipe within it can incur severe bending and consequent tensile failure.
- Seabed gouging by ice: In freezing waters, floating ice features often drift into shallower waters, and their keel comes into contact with the seabed. As they continue to drift, they gouge the seabed to a given depth. Pipeline route planning for these areas has to consider laying the pipeline in a trench with enough room above it to avoid 1) a direct hit, and 2) excessive bending as a result of gouging-induced soil displacement.
- Currents: High currents are objectionable in that they hinder pipe laying operations. For instance, in shallow seas tidal currents may be quite strong in a straight between two islands. Under these circumstances, it may be preferable to bring the pipe elsewhere, even if this alternative route ends up being longer.
- Waves: In shallow waters, waves can also be problematic for pipeline laying operations (in severe wave regimes) and, subsequently, to its stability, because of the water's scouring action. This is one of a number of reasons why landfalls (where the pipeline reaches the shoreline) are particularly delicate areas to plan.

OTHER USES OF THE SEABED

Proper planning of a pipeline route has to factor in a wide range of human activities that make use of the seabed along the proposed route, or that are likely to do so in the future.

They include the following:

- Other pipelines: If and where the proposed pipeline intersects an existing one, which is not uncommon, a bridging structure may be required at that juncture in order to cross it. This has to be done at a right angle. The juncture should be carefully designed so as to avoid interferences between the two structures either directly through mechanical contact or indirectly because of hydrodynamic forces.
- Fishing vessels: Commercial fishing makes use of heavy fishing nets dragged on the seabed and extending several kilometers behind the trawler. This net could snag the pipeline, with potential damage to both pipeline and vessel.
- Ship anchors: Ship anchors are a potential threat to pipelines, especially near harbors.
- Military activities: Some areas still have mines originating from former conflicts but that are still operational. Other areas, used for bombing or gunning practices, may also conceal live ammunition. Moreover, at some locations, various types of instrumentation are laid on the seafloor for submarine detection. These areas have to be avoided.

SUBSEA PIPELINE CHARACTERISTICS

Submarine pipelines generally vary in diameter from 3 inches (76 mm) for gas lines, to 72 inches (1,800 mm) for high capacity lines. Wall thicknesses typically range from 10 millimetres (0.39 in) to 75 millimetres (3.0 in). The pipe can be designed for fluids at high temperature and pressure. The walls are made from high-yield strength steel, 350-500 MPa (50,000-70,000 psi), weldability being one of the main selection criteria.

The structure is often shielded against external corrosion by coatings such as bitumastic or epoxy, supplemented by cathodic protection with sacrificial anodes. Concrete or fiberglass wrapping provides further protection against abrasion. The addition of a concrete coating is also useful to compensate for the pipeline's negative buoyancy when it carries lower density substances. The pipeline's inside wall is not coated for petroleum service. But when it carries seawater or corrosive substances, it can be coated with epoxy, polyurethane or polyethylene; it can also be cement-lined.

In the petroleum industry, where leaks are unacceptable and the pipelines are subject to internal pressures typically in the order of 10 MPa (1500 psi), the segments are joined by full penetration welds. Mechanical joints are also used. A pig is a standard device in pipeline transport, be it on-land or offshore. It is used to test for hydrostatic pressure, to check for dents and crimps on the sidewalls inside the pipe, and to conduct periodic cleaning and minor repairs.

PIPELINE CONSTRUCTION

Pipeline construction involves two procedures: assembling a large number of pipe segments into a full line, and installing that line along the desired route. Several systems can be used - for a submarine pipeline, the choice in favor of any one of them is based on the following factors: physical and environmental conditions (e.g. currents, wave regime), availability of equipment and costs, water depth, pipeline length and diameter, constraints tied to the presence of other lines and structures along the route. These systems are generally divided into four broad categories: pull/tow, S-lay, J-lay and reel-lay

THE PULL/TOW SYSTEM

In the pull/tow system, the submarine pipeline is assembled onshore and then towed to location. Assembly is done either parallel or perpendicular to the shoreline - in the former case, the full line can be built prior to tow out and installation. A significant advantage with this system is that pre-testing and inspection of the line are done onshore, not at sea. It allows to handle lines of any size and complexity. As for the towing procedures, a number of configurations can be used, which may be categorized as follows: surface tow, near-surface tow, mid-depth tow and off-bottom tow.

- Surface tow: In this configuration, the pipeline remains at the surface of the water during tow, and is then sunk into position at lay site. The line has to be buoyant - this can be done with individual buoyancy units attached to it. Surface tows are not appropriate for rough seas and are vulnerable to lateral currents.
- Near-surface tow: The pipeline remains below the water surface but close to it - this mitigates wave action. But the spar buoys used to maintain the line at that level are affected by rough seas, which in itself may represent a challenge for the towing operation.
- Mid-depth tow: The pipeline is not buoyant - either because it is heavy or it is weighted down by hanging chains. In this configuration, the line is suspended in a catenary between two towing vessels. The shape of that catenary (the sag) is a balance between the line's weight, the tension applied to it by the vessels and hydrodynamic lift on the chains. The amount of allowable sag is limited by how far down the seabed is.
- Off-bottom tow: This configuration is similar to the mid-depth tow, but here the line is maintained within 1 to 2 m (several feet) away from the bottom, using chains dragging on the seabed.
- Bottom tow: In this case, the pipeline is dragged onto the bottom - the line is not affected by waves and currents, and if the sea gets too rough for the tow vessel, the line can simply be abandoned and recovered later. Challenges with this type of system include:

requirement for an abrasion-resistant coating, interaction with other submarine pipelines and potential obstructions (reef, boulders, etc.). Bottom tow is commonly used for river crossings and crossings between shores.

THE S-LAY SYSTEM

In the S-lay system, the pipeline assembly is done at the installation site, on-board a vessel that has all the equipment required for joining the pipe segments: pipe handling conveyors, welding stations, X-ray equipment, joint-coating module, etc. The S notation refers to the shape of the pipeline as it is laid onto the seabed. The pipeline leaves the vessel at the stern from a supporting structure called a stringer that guides the pipe's downward motion and controls the convex-upward curve (the overbend).

As it continues toward the seabed, the pipe has a convex-downward curve (the sagbend) before coming into contact with the seabed (touch down point). The sagbend is controlled by a tension applied from the vessel (via tensioners) in response to the pipeline's submerged weight. The pipeline configuration is monitored so that it will not get damaged by excessive bending.

This on-site pipeline assembly approach, referred to as lay-barge construction, is known for its versatility and self-contained nature - despite the high costs associated with this vessel's deployment, it is efficient and requires relatively little external support. But it may have to contend with severe sea states - these adversely affect operations such as pipe transfer from supply boats, anchor-handling and pipe welding. Recent developments in lay-barge design include dynamic positioning and the J-lay system.

THE J-LAY SYSTEM

In areas where the water is very deep, the S-lay system may not be appropriate because the pipeline leaves the stinger to go almost straight down. To avoid sharp bending at the end of it and to mitigate excessive sag bending, the tension in the pipeline would have to be high. Doing so would interfere with the vessel's positioning, and the tensioner could damage the pipeline. A particularly long stinger could be used, but this is also objectionable since that structure would be adversely affected by winds and currents. The J-lay system, one of the latest generations of lay-barge, is better suited for deep water environments. In this system, the pipeline leaves the vessel on a nearly vertical ramp (or tower). There is no overbend - only a sagbend of catenary nature (hence the J notation), such that the tension can be reduced. The pipeline is also less exposed to wave action as it enters the water. However, unlike for the S-lay system, where pipe welding can be done simultaneously at several locations along the vessel deck's length, the J-lay system can only accommodate one welding station. Advanced methods of automatic welding are used to compensate for this drawback.

THE REEL-LAY SYSTEM

In the reel-lay system, the pipeline is assembled onshore and is spooled onto a large drum typically about 20 metres (66 ft) x 6 metres (20 ft) in size, mounted onboard a purpose-built vessel. The vessel then goes out to location to lay the pipeline. Onshore facilities to assemble the pipeline have inherent advantages: they are not affected by the weather or the sea state and are less expensive than seaborne operations.

Pipeline supply can be coordinated: while one line is being laid at sea, another one can be spooled onshore. A single reel can have enough capacity for a full length flow line. The reel-lay system, however, can only handle lower diameter pipelines - up to about 400 mm (16 in). Also, the kind of steel making up the pipes must be able to undergo the required amount of plastic deformation as it is bent to proper curvature (by a spiral J-tube) when reeled around the drum, and straightened back (by a straightener) during the layout operations at the installation site.

OFFSHORE GEOTECHNICAL ENGINEERING

Offshore geotechnical engineering is a sub-field of geotechnical engineering. It is concerned with foundation design, construction, maintenance and de-commissionning for human-made structures in the sea. Oil platforms, artificial islands and submarine pipelines are examples of such structures. The seabed has to be able to withstand the weight of these structures and the applied loads.

Geohazards must also be taken into account. The need for offshore developments stems from a gradual depletion of hydrocarbon reserves onshore or near the coastlines, as new fields are being developed at greater distances offshore and in deeper water, with a corresponding adaptation of the offshore site investigations. Today, there are more than 7,000 offshore platforms operating at a water depth up to and exceeding 2000 m. A typical field development extends over tens of square kilometers, and may comprise several fixed structures, infield flowlines with an export pipeline either to the shoreline or connected to a regional trunkline.

DIFFERENCES BETWEEN ONSHORE AND OFFSHORE GEOTECHNICAL ENGINEERING

An offshore environment has several implications for geotechnical engineering. These include the following:

- Ground improvement (on the seabed) and site investigation are expensive.
- Soil conditions are unusual (e.g. presence of carbonates, shallow gas).
- Offshore structures are tall, often extending over 100 metres (330 ft) above their foundation.
- Offshore structures typically have to contend with significant lateral

loads (i.e. large moment loading relative to the weight of the structure).

- Cyclic loading can be a major design issue.
- Offshore structures are exposed to a wider range of geohazards.
- The codes and technical standards are different from those used for onshore developments.
- Design focuses on ultimate limit state as opposed to deformation.
- Design modifications during construction are either unfeasible or very expensive.
- The design life of these structures often ranges between 25-50 years.
- The environmental and financial costs in case of failure can be higher.

THE OFFSHORE ENVIRONMENT

Offshore structures are exposed to various environmental loads: wind, waves, currents and, in cold oceans, sea ice and icebergs. Environmental loads act primarily in the horizontal direction, but also have a vertical component. Some of these loads get transmitted to the foundation (the seabed).

Wind, wave and current regimes can be estimated from meteorological and oceanographic data, which are collectively referred to as metocean data. Earthquake-induced loading can also occur - they proceed in the opposite direction: from the foundation to the structure. Depending on location, other geohazards may also be an issue. All of these phenomena may affect the integrity or the serviceability of the structure and its foundation during its operational lifespan - they need to be taken into account in offshore design.

THE NATURE OF THE SOIL

Following are some to the features characterizing the soil in an offshore environment:

- The soil is made up of sediments, which are generally assumed to be in a saturated state - saline water fills in the pore space.
- Marine sediments are composed of detrital material as well as remains of marine organisms, the latter making up calcareous soils.
- Total sediment thickness varies on a regional scale - it is normally higher near the coastline than it is away from it, where it is also finer grained.
- In places, the seabed can be devoid of sediment, due to strong bottom currents.
- The consolidation state of the soil is either normally consolidated (due to slow sediment deposition), overconsolidated (in places, a relic of glaciation) or underconsolidated (due to high sediment input).

METOCEAN ASPECTS

Wave forces induce motion of floating structures in all six degrees of freedom - they are a major design criterion for offshore structures. When a wave's orbital motion reaches the seabed, it induces sediment transport. This only occurs to a water depth of about 200 metres (660 ft), which is the commonly adopted boundary between shallow water and deep water. In shallow water, waves may generate pore pressure build-up in the soil, which may lead to flow slide, and repeated impact on a platform may cause liquefaction and loss of support.

Currents are a source of horizontal loading for offshore structures. Because of the Bernoulli effect, they may also exert upward or downward forces on structural surfaces and can induce the vibration of wire lines and pipelines. Currents are responsible for eddies around a structure, which cause scouring and erosion of the soil. There are various types of currents: oceanic circulation, geostrophic, tidal, wind-driven, and density currents.

GEOHAZARDS

Geohazards are associated with geological activity, geotechnical features and environmental conditions. Shallow geohazards are those occurring at less than 400 metres (1,300 ft) below the seafloor. Information on the potential risks associated with these phenomena is acquired through studies of the geomorphology, geological setting and tectonic framework in the area of interest, as well as with geophysical and geotechnical surveys of the seafloor. Examples of potential threats include tsunamis, landslides, active faults, mud diapirs and the nature of the soil layering (presence of karst, gas hydrates, carbonates). In cold regions, gouging ice features are a threat to subsea installations, such as pipelines. The risks associated with a particular type of geohazard is a function of how exposed the structure is to the event, how severe this event is and how often it occurs (for episodic events). Any threat has to be monitored, and mitigated for or removed.

SITE INVESTIGATION

Offshore site investigations are not unlike those conducted onshore. They may be divided into three phases:

- A desk study, which includes data compilation.
- Geophysical surveys, either shallow and deep seabed penetration.
- Geotechnical surveys, which includes sampling/drilling and in situ testing.

DESK STUDY

In this phase, which may take place over a period of several months (depending on project size), information is gathered from various sources, including reports, papers and databases, with the purpose of evaluating risks, assessing design options and planning the subsequent phases. Bathymetry, regional geology, potential geohazards, seabed obstacles and metocean data. are some of the information that are sought after during that phase.

GEOPHYSICAL SURVEYS

Geophysical surveys can be used for various purposes. One is to study the bathymetry in the location of interest and to produce an image of the seafloor (irregularities, objects on the seabed, lateral variability, ice gouges, ...). Seismic refraction surveys can be done to obtain information on shallow seabed stratigraphy - it can also be used to locate material such as sand and gravel for use in the construction of artificial islands. Geophysical surveys are conducted from a research vessel equipped with sonar devices and related equipment, such as single-beam and multibeam echosounders, side-scan sonars, 'towfish' and remotely operated vehicles (ROVs). For the sub-bottom stratigraphy, the tools used include boomers, sparkers, pingers and chirp. Geophysical surveys are normally required before conducting the geotechnical surveys; in larger projects, these phases may be interwoven.

GEOTECHNICAL SURVEYS

Geotechnical surveys involve a combination of sampling, drilling, in situ testing as well as laboratory soil testing both onshore and offshore. They serve to ground truth the results of the geophysical investigations; they also provide a detailed account of the seabed stratigraphy and soil engineering properties. Depending on water depth and metocean conditions, geotechnical surveys may be conducted from a dedicated geotechnical drillship, a semi-submersible, a jackup rig, a large hovercraft or other means. They are done at a series of specific locations, while the vessel maintains a constant position. Dynamic positioning and mooring with four-point anchoring systems are used for that purpose.

Shallow penetration geotechnical surveys may include soil sampling of the seabed surface or in situ mechanical testing. They are used to generate information on the physical and mechanical properties of the seabed. They extend to the first few meters below the mudline. Surveys done to these depths, which may be conducted at the same time as the shallow geophysical survey, may suffice if the structure to be deployed at that location is relatively light. These surveys are also useful for planning subsea pipeline routes.

The purpose of deep penetration geotechnical surveys is to collect information on the seabed stratigraphy to depths extending up to a few 100 meters below the mudline. These surveys are done when larger structures are planned at these locations. Deep drill holes make require a few days during which the drilling unit has to remain exactly in the same position.

SAMPLING AND DRILLING

Seabed surface sampling can be done with a grab sampler and with a box corer. The latter provides undisturbed specimens, on which testing can be conducted, for instance, to determine the soil's relative density, water content and mechanical properties. Sampling can also be achieved with a tube corer, either gravity-driven, or that can be pushed into the seabed by a piston or by means of a vibration system (a device called a vibrocorer). Drilling is another means of sampling the seabed. It is used to obtain a record of the seabed stratigraphy or the rock formations below it. The set-up used to sample an offshore structure's foundation is similar to that used by the oil industry to reach and delineate hydrocarbon reservoirs, with some differences in the types of testing. The drill string consists of a series of pipe segments 5 inches (13 cm) in diameter screwed end to end, with a drillbit assembly at the bottom. As the dragbit (teeth extending downward from the drillbit) cut into the soil, soil cuttings are produced. Viscous drilling mud flowing down the drillpipe collects these cuttings and carry them up outside the drillpipe. As is the case for onshore geotechnical surveys, different tools can be used for sampling the soil from a drill hole, notably "Shelby tubes", "piston samplers" and "split spoon samplers".

IN SITU SOIL TESTING

Information on the mechanical strength of the soil can be obtained in situ (from the seabed itself as opposed to in a laboratory from a soil sample). The advantage of this approach is that the data are obtained from soil that has not suffered any disturbance as a result of its relocation. Two of the most commonly used instruments used for that purpose are the cone penetrometer (CPT) and the shear vane. The CPT is a rod-shaped tool whose end has the shape of a cone with a known apex angle (e.g. 60 degrees). As it is pushed into the soil, the resistance to penetration is measured, thereby providing an indication of soil strength. A sleeve behind the cone allows the independent determination of the frictional resistance. Some cones are also able to measure pore water pressure. The shear vane test is used to determine the undrained shear strength of soft to medium cohesive soils. This instrument usually consists of four plates welded at 90 degrees from each other at the end of a rod. The rod is then inserted into the soil and a torque is applied to it so as to achieve a constant rotation rate. The torque resistance is measured and an equation is then used to determine the undrained shear strength (and the residual strength), which takes into account the vane's size and geometry.

OFFSHORE STRUCTURES AND GEOTECHNICAL CONSIDERATIONS

Offshore structures are mainly represented by platforms, notably jackup rigs, steel jacket structures and gravity-based structures. The nature of the seabed has to be taken into account when planning these developments. For instance, a gravity-based structure typically has a very large footprint and is relatively buoyant (because it encloses a large open volume). Under these circumstances, vertical loading of the foundation may not be as significant as the horizontal loads exerted by wave actions and transferred to the seabed.

In that scenario, sliding could be the dominant mode of failure. A more specific example is that of the Woodside "North Rankin A" steel jacket structure offshore Australia. The shaft capacity for the piles making up each of the structure's legs was estimated on the basis of conventional design methods, notably when driven into siliceous sands. But the soil at that site was a lower capacity calcareous sand. Costly remediation measures were required to correct this oversight.

Proper seabed characterization is also required for mooring systems. For instance, the design and installation of suction piles has to take into account the soil properties, notably its undrained shear strength. The same is true for the installation and capacity assessment of plate anchors.

Subsea pipelines are another common type of man-made structure in the offshore environment. These structures either rest on the seabed, or are placed inside a trench to protect them from fishing trawlers, dragging anchors or fatigue due current-induced oscillations. Trenching is also used to protect pipelines from gouging by ice keels. In both cases, planning of the pipeline involves geotechnical considerations. Pipelines resting on the seabed require geotechnical data along the proposed pipeline route to evaluate potential stability issues, such as passive failure of the soil below it (the pipeline drops) due to insufficient bearing capacity, or sliding failure (the pipeline shift sideways), due to low sliding resistance. The process of trenching, when required, needs to take into account soil properties and how they would affect ploughing duration. Buckling potential induced by the axial and transverse response of the buried pipeline during its operational lifespan need to be assessed at the planning phase, and this will depend on the resistance of the enclosing soil.

15

Beach Engineering

BEACH RIDGE

A beach ridge is a wave-swept or wave-deposited ridge running parallel to a shoreline. It is commonly composed of sand as well as sediment worked from underlying beach material. The movement of sediment by wave action is called *littoral transport*. Movement of material parallel to the shoreline is called *longshore transport*. Movement perpendicular to the shore is called *on-offshore transport*. A beach ridge may be capped by, or associated with, sand dunes. The height of a beach ridge is affected by wave size and energy. A fall in water level (or an uplift of land) can isolate a beach ridge from the body of water that created it.

Isolated beach ridges may be found along dry lakes in the western United States and inland of the Great Lakes of North America, where they formed at the end of the last ice age when lake levels were much higher due to glacial melting and obstructed outflow caused by glacial ice. Some isolated beach ridges are found in parts of Scandinavia, where glacial melting relieved pressure on land masses and resulted in subsequent crustal lifting or post-glacial rebound. A rise in water level can submerge beach ridges created at an earlier stage, causing them to erode and become less distinct. Beach ridges may become routes for roads and trails.

BEACHROCK

Beachrock is a friable to well-cemented sedimentary rock that consists of a variable mixture of gravel-, sand-, and silt-sized sediment that is cemented with carbonate minerals and has formed along a shoreline. Depending on location, the sediment that is cemented to form beachrock can consist of a variable mixture of shells, coral fragments, rock fragments of different types, and other materials.

It can contain scattered artifacts, pieces of wood, and coconuts. Beachrock typically forms within the intertidal zone within tropical or semitropical regions. However, Quaternary beachrock is also found as far north and south as 60' latitude.

OVERVIEW

Beachrock units form under a thin cover of sediment and generally overlie unconsolidated sand. They typically consist of multiple units, representing multiple episodes of cementation and exposure. The mineralogy of beachrocks is mainly high-magnesium calcite or aragonite. The main processes involved in the cementation are : supersaturation with $CaCO_3$ through direct evaporation of seawater (Scoffin, 1970), groundwater CO_2 degassing in the vadose zone (Hanor, 1978), mixing of marine and meteoric water fluxes (Schmalz, 1971) and precipitation of micritic calcium carbonate as a byproduct of microbiological activity (Neumeier, 1999). On retreating coasts, outcrops of beachrock may be evident offshore where they may act as a barrier against coastal erosion.

Beachrock presence can also induce sediment deficiency in a beach and out-synch its wave regime. Because beachrock is lithified within the intertidal zone and because it commonly forms in a few years, its potential as an indicator of past sea level is important.

CEMENTATION AND POSITION OF BEACHROCK

Beachrocks are located along the coastline in a parallel term and they are usually a few meters offshore. They are generally separated in several levels which may correspond to different generations of beachrock cementation. Thus, the older zones are located in the outer part of the formation when the younger ones are on the side of the beach, possibly under the unconsolidated sand. They also seem to have a general inclination to the sea (50 – 150). There are several appearances of beachrock formations which are characterized by multiple cracks and gaps. The result from this fact is an interruptible formation of separated blocks of beachrock, which may be of the same formation.

The length of beachrocks varies from meters to kilometers, its width can reach up to 300 meters and its height starts from 30 cm and reaches 3 meters. Following the process of coastal erosion, beachrock formation may be uncovered. Coastal erosion may be the result of sea level rise or deficit in sedimentary equilibrium. One way or another, unconsolidated sand that covers the beachrock draws away and the formation is revealed. If the process of cementation continues, new beachrock would be formed in a new position in the intertidal zone. Successive phases of sea level change may result in sequential zones of beachrock.

POCKET BEACH

Pocket beach is usually a small beach, between two headlands. In an idealized setting, there is very little or no exchange of sediment between the pocket beach and the adjacent shorelines. Pocket beaches can be natural or artificial. Many natural pocket beaches exist throughout the world. Artificial pocket beaches are usually constructed in areas where natural beaches are fairly narrow or absent. Examples of artificial pocket beaches include over 100 such systems on Chesapeake Bay in the United States, each consisting of several individual pocket beaches; the Fisher Key, Florida project constructed on a dredge spoil island originally consisting of cobble dredge material;, and the Fred Howard Park Beach that was constructed offshore of a muddy mangrove shoreline.

Additionally, there have been many pocket beaches constructed in the Caribbean where resorts have been developed along rocky shorelines with minimal natural beaches.

SHINGLE BEACH

A shingle beach is a beach which is armoured with pebbles or small- to medium-sized cobbles (as opposed to fine sand). Typically, the stone composition may grade from characteristic sizes ranging from two to 200 mm diameter.

While this beach landform is most commonly found in Western Europe, examples are found in Bahrain, North America and in a number of other world regions, such as the east coast of New Zealand's South Island, where they are associated with the shingle fans of braided rivers. Though created at shorelines, post-glacial rebound can raise shingle beaches as high as 200 metres above sea level, at the High Coast in Sweden. The ecosystems formed by this unique association of rock and sand allow colonization by a variety of rare and endangered species.

FORMATION

Shingle beaches are typically steep, because the waves easily flow through the coarse, porous surface of the beach, decreasing the effect of backwash erosion and increasing the formation of sediment into a steeply sloping beach.

TOURISM

Shingle beaches are often criticized as undesirable for visitors. Canterbury City Council notes that the nearby shingle beach at Whitstable is uncomfortable to walk and lie on. Also, advertisers have been known to replace images of shingle beaches with sand in promotional material.

Beach morphodynamics

Beach morphodynamics refers to the study of the interaction and adjustment of the seafloor topography and fluid hydrodynamic processes, seafloor morphologies and sequences of change dynamics involving the motion of sediment. Hydrodynamic processes include those of waves, tides and wind-induced currents.

While hydrodynamic processes respond instantaneously to morphological change, morphological change requires the redistribution of sediment. As sediment takes a finite time to move, there is a lag in the morphological response to hydrodynamic forcing. Sediment can therefore be considered to be a time-dependent coupling mechanism. Since the boundary conditions of hydrodynamic forcing change regularly, this may mean that the beach never attains equilibrium.

This systems approach to the coast was developed by Wright and Thom in 1977. Examples of beach morphodynamic processes include the formation of beach cusps (on a very small scale), intertidal bars and log-spiral (Yasso 1965)/crenulate (Silvester 1980) embayments. Morphodynamic processes exhibit positive and negative feedbacks (such that beaches can, over different timescales, be considered to be both self-forcing and self-organised systems), nonlinearities and threshold behaviour. According to their dynamic and morphological characteristics, exposed sandy beaches can be classified into several morphodynamic types (Short & Wright 1983, Short 1996). There is a large scale of morphodynamic states, this scale ranges from the dissipative state to the reflective extremes.

Dissipative beaches are flat, have fine sand, incorporating waves that tend to break far from the intertidal zone and dissipate force progressively along wide surf zones. Reflective beaches are steep, and are known for their coarse sand; they have no surf zone, and the waves break brusquely on the intertidal zone. Depending on beach state, near bottom currents show variations in the relative dominance of motions due to: incident waves, subharmonic oscillations, infragravity oscillations, and mean longshore and rip currents. On reflective beaches, incident waves and subharmonic edge waves are dominant.

In highly dissipative surf zones, shoreward decay of incident waves is accompanied by shoreward growth of infragravity energy; in the inner surf zone, currents associated with infragravity standing waves dominate. On intermediate states with pronounced bar-trough (straight or crescentic) topographies, incident wave orbital velocities are generally dominant but significant roles are also played by subharmonic and infragravity standing waves, longshore currents, and rips. The strongest rips and associated feeder currents occur in association with intermediate transverse bar and rip topographies.(Virginia Institute of Marine Science- Scopus 460).

With respect to beach morphodynamics, a spectrum of morphodynamic states has been developed, ranging from reflective beaches at one end to dissipative beaches at the other. Reflective beaches are typically steep in profile with a narrow shoaling and surf zone, composed of coarse sediment, and characterised by surging breakers. Coarser sediment allows percolation during the swash part of the wave cycle, thus reducing the strength of backwash and allowing material to be deposited in the swash zone. At the other end of the continuum of beach states, dissipative beaches are wide and flat in profile, with a wide shoaling and surf zone, composed of finer sediment, and characterised by spilling breakers. Transitions between beach states are often caused by changes in wave energy, with storms causing reflective beach profiles to flatten (offshore movement of sediment under steeper waves), thus adopting a more dissipative profile.

Morphodynamic poos are also associated with other coastal landforms, for example spur and groove topography on coral reefs and tidal flats in infilling estuaries.

16

Coastal Hazards

Coastal Hazards are both natural and man-made disasters that happen along the coastline. This article will discuss the coastal environments on a global scale, as well as the causes of coastal hazards (ex. hurricanes and nor'easter) in which the environments are affected. Coastal policies and management and planning are later discussed to provide adequate information on implementation and mitigation of these coastal hazards.

In Coastal hazards play a major role in today's society because it is a part of human nature to live near or along the coast. 80% of people live near the coast. 1.2 billion people live within 100 km of the coast and it is on the rise. It is important for us to educate ourselves and others on coastal hazards so we can continue living near the coast with the least amount of damage to the environment. In the past, human development has effected our coastal living arrangements by making it vulnerable to such fragile environments such as the barrier islands.

Disasters such as hurricanes, with high winds and swells, cause erosion along coastlines. Due to this, certain policies have been set in place to try to manage disaster property damages; FEMA, NFIP, CZM. Adaptive management has become a major source of planning in order to make development sustainable for the environment. Structural versus non-structural mitigation techniques have been a main focus for planners towards coastal hazards. Short term solutions versus long term solutions; dune, sea walls, etc., which one should planners focus on and overall which would be the better option to spend funds on. Current strategies are only an illusion of safety towards our coastlines.

For coastal hazards, it is important to have emergency management plans in advance so agencies can respond quickly and effectively. If they have better plans it will create a faster recovery time that will ensure economic, social, and environmental life back to its original state. For example, Hurricane Katrina was a national eye opener on how on the risk of improper risk management between federal, state, and local governments. Due to the nature of the federal system and policies, often conflict with one another between both private and public agencies in addition to mandated mitigation programs going unfunded. Communication between these governments is essential in coordinating an effective response through mitigation. Coastal hazards are unpredictable to know where and when they will occur and it is important for everyone to be prepared.

COASTAL ENVIRONMENTS

There are many different types of environments along the coasts of the United States with very diverse features that affect, influence, and mold the near-shore processes that are involved. Understanding these ecosystems and environments can further advance the mitigating techniques and policy-making efforts against natural and man-made coastal hazards in these vulnerable areas.

The five most common types of coastal zones range from the northern ice-pushing, mountainous coastline of Alaska and Maine, the barrier island coasts facing the Atlantic, the steep, cliff-back headlands along the pacific coast, the marginal-sea type coastline of the Gulf region, and the coral reef coasts bordering Southern Florida and Hawaii.

ICE-PUSHING/MOUNTAINOUS COASTLINE

These coastal regions along the northernmost part of the nation were affected predominantly by, along with the rest of the Pacific Coast, continuous tectonic activity, forming a very long, irregular, ridged, steep and mostly mountainous coastline. These environments are heavily occupied with permafrost and glaciers, which are the two major conditions affecting Alaska's Coastal Development.

BARRIER ISLAND COASTLINE

Barrier islands are a land form system that consists of fairly narrow strips of sand running parallel to the mainland and play a significant role in mitigating storm surges and oceans swells as natural storm events occur. The morphology of the various types and sizes of barrier islands depend on the wave energy, tidal range, basement controls, and sea level trends. The islands create multiple unique environments of wetland systems including marshes, estuaries, and lagoons.

STEEP, CLIFF-BACKING ABRASION COASTLINE

The coastline along the western part of the nation consists of very steep, cliffed rock formations generally with vegetative slopes descending down and a fringing beach below.

The various sedimentary, metamorphic, and volcanic rock formations assembled along a tectonically disturbed environment, all with altering resistances running perpendicular, cause the ridged, extensive stretch of uplifted cliffs that form the peninsulas, lagoons, and valleys.

Marginal-sea type coastline

The southern banks of the United States border the Gulf of Mexico, intersecting numerous rivers, forming many inlets bays, and lagoons along its coast, consisting of vast areas of marsh and wetlands. This region of landform is prone to natural disasters yet highly and continuously developed, with man-made structures attaining to water flow and control.

Coral reef coastline

Coral reefs are located off the shores of the southern Florida and Hawaii consisting of rough and complex natural structures along the bottom of the ocean floor with extremely diverse ecosystems, absorbing up to ninety percent of the energy dissipated from wind-generated waves.

This process is a significant buffer for the inner-lying coastlines, naturally protecting and minimizing the impact of storm surge and direct wave damage. Because of the highly diverse ecosystems, these coral reefs not only provide for the shoreline protection, but also deliver an abundant amount of services to fisheries and tourism, increasing its economic value.

CAUSES OF COASTAL HAZARDS

NATURAL VS HUMAN DISASTERS

The population that lives along or near our coastlines are an extremely vulnerable population. There are numerous issues facing our coastlines and there are two main categories that these hazards can be placed under, National disasters and Human disasters. Both of these issues cause great damage to our coastlines and discussion is still ongoing regarding what standards or responses need to be met to help both the individuals who want to continue living along the coastline, while keeping them safe and not eroding more coastline away. Natural disasters are disasters that are out of human control and are usually caused by the weather.

Disasters that include but are not limited to; storms, tsunamis, typhoons, flooding, tides, waterspouts, nor'easters, and storm surge. Human disasters occur when humans are the main culprit behind why the disaster happened. Some human disasters are but are not limited to; pollution, trawling, and human development. Natural and human disasters continue to harm the coastlines severely and they need to be researched in order to prepare/stop the hazards if possible.

The populations that live near or along the coast experience many hazards and it affects millions of people. Around ten million people globally feel the effects of coastal problems yearly and most are due to certain natural hazards like coastal flooding with storm surges and typhoons. A major problem related to coastal regions deals with how the entire global environment is changing and in response, the coastal regions are easily effected.

STORMS, FLOODING, EROSION

Storms are one of the major hazards that are associated to coastal regions. Storms, flooding, and erosion are closely associated and can happen simultaneously. Tropical storms or Hurricanes especially can devastate coastal regions. For example, Florida during Hurricane Andrew occurred in 1992 that caused extreme damage.

It was a category five hurricane that caused $26.5 billion in damages and even 23 individuals lost their lives from the storm. Hurricane Katrina also caused havoc along the coast to show the extreme force a hurricane can do in a certain region. In almost all cases, storms are the major culprit that causes flooding and erosion. Flash flooding is caused by storms that occurs when a massive amount of rainfall comes down into an area over a short period of time. Where as a storm surge, which is closely related to tropical storms, is when the wind collects and pushes water towards low pressure or inland and can rise rapidly. It is an offshore rise of water and overall creates a higher sea level that rises and is pushed inland. The amount of rise or fall of storm surge depends greatly on the amount and duration of wind and water in a specific location. Also if it occurs during a high tide it can have an even greater effect on the coast.

Almost all storms with high wind and water cause erosion along the coast. Erosion occurs when but not limited to; along shore currents, tides, sea level rise and fall, and high winds. Larger amounts of erosion cause the coastline to erode away at a faster rate and can leave people homeless and leave less land to develop or keep for environmental reasons. Coastal erosion has been increasing over the past few years and it is still on the rise which makes it a major coastline hazard. In the United States, 45 percent of its coast line is along the Atlantic or Gulf coast and the erosion rate per year along the Gulf coast is at six feet a year. The average rate of erosion along the Atlantic is around two to three feet a year. Even with these findings, erosion rates in specific locations vary because of various environmental factors such as major storms that can cause major erosion upwards to 100 feet or more in only one day.

POLLUTION, TRAWLING, HUMAN DEVELOPMENT

Pollution, trawling, and human development are major human disasters that effect coastal regions. There are two main categories related to pollution, point source pollution, and nonpoint source pollution. Point source pollution is when there is an exact location such as a pipeline or a body of water that leads into the rivers and oceans. Known dumping into the ocean is also another point source of pollution. Nonpoint source pollution would pertain more to fertilizer runoff, and industrial waste. Examples of pollution that effect the coastal regions are but are not limited to; fertilizer runoff, oil spills, and dumping of hazardous materials into the oceans. More human acts that hurt the coastline are as follows; waste discharge, fishing, dredging, mining, and drilling. Oil spills are one of the most hazardous dangers towards coastal communities. They are hard to contain, difficult to clean up, and devastate everything. The fish, animals such as birds, the water, and especially the coastline near the spill. The most recent oil spill that had everybody concerned with oil spill was the BP oil spill.

Trawling hurts the normal ecosystems in the water around the coastline. It depletes all ecosystems on the ocean floor such as, flounder, shellfish, marsh etc.. It is simply a giant net that is drug across the ocean floor and destroys and catches anything in its path. Human development is one of the major problems when facing coastal hazards. The overall construction of buildings and houses on the coast line takes away the natural occurrences to handle the fluctuation in water and sea level rise. Building houses in pre-flood areas or high risk areas that are extremely vulnerable to flooding are major concerns towards human development in coastal regions. Having houses and buildings in areas that are known to have powerful storms that will create people to be in risk by living there. Also pertaining to barrier islands, where land is at risk for erosion but they still continue to build there anyway. More and more houses today are being taken by the ocean, look at picture above.

POLICIES

NATIONAL FLOOD INSURANCE PROGRAM

The National Flood Insurance Program or NFIP was instituted in 1968 and offers home owners in qualifying communities an opportunity to rebuild and recover after flooding events following the decision by insurance companies to discontinue providing flood insurance. This decision was made on behalf of the private insurers after continually high and widespread flood losses. The goals of this program are to not only better protect individuals from flood, but to reduce property losses, and reduce the total amount disbursed for flood loses by the government. Only communities which have adopted and implemented mitigation policies that are compliant with or exceed federal regulations. The regulatory policies reduce risk to life and property located within floodplains. The NFIP also comprehensively mapped domestic floodplains increasing public awareness of risk. The majority of structures were constructed after the mapping was completed and risk could be assessed. To reduce the cost to these owners, which consititue roughly 25% of the total policies the rates for insurance are subsidized.

COASTAL STATES ORGANIZATION

The Coastal States Organization or CSO was established in 1970 to represent 35 U.S. sub-federal governments on issues of coastal policies. CSO lobbies Congress on issues pertaining to Coastal Policy allowing states input on federal policy decisions. Funding, support, water quality, coastal hazards, and coastal zone management are the primary issues CSO promotes. The strategic goals of CSO are to provide information and assistance to members,evaluate and manage coastal needs, and secure long term funding for member states initiatives.

COASTAL ZONE MANAGEMENT ACT

In 1972 the Coastal Zone Management Act or CZMA works to streamline the policies which states create to a minimum federal standard for environmental protection. CZMA establishes the national policy for the development and implementation of regulatory programs for coastal land usage, which is supposed to be reflected in state legislation such as CAMA. CZMA also provides minimum building requirements to make the insurance provided through the NFIP less expensive for the government to operate by mitigating losses. Congress found that it was necessary to establish the minimum which programs should provide for. Each coastal state is required to have a program with 7 distinct parts: Identifying land uses,Identifying critical coastal areas, Management measures,Technical assistance, Public participation, Administrative coordination, State coastal zone boundary modification.

THE COASTAL AREA MANAGEMENT ACT

The Coastal Area Management Act or CAMA is policy that was implemented by the state of North Carolina in 1974 to work in-tandem with the CZMA. It creates a cooperative program between the state and local governments. The State government operates in an advisory capacity and reviews decisions made by local government planners. The goal of this legislation was to create a management system capable of preserving the coastal environment, insure the preservation of land and water resources, balance the use of coastal resources and establish guidelines and standards for conservations, economic development, tourism, transportation, and the protection of common law.

MANAGEMENT AND PLANNING

Due to the increasing urbanization along the coastlines, planning and management are essential to protecting the ecosystems and environment from depleting. Coastal management is becoming implemented more because of the movement of people to the shore and the hazards that come with the territory. Some of the hazards include movement of barrier islands, sea level rise, hurricanes, nor'easters, earthquakes, flooding, erosion, pollution and human development along the coast. The Coastal Zone Management Act (CZMA) was created in 1972 because of the continued growth along the coast, this act introduced better management practices such as integrated coastal zone management, adaptive management and the use mitigation strategies when planning. According to the Coastal Zone Management Act, the objectives are to remain balanced to "preserve, protect, develop, and where possible, to restore or enhance the resources of the nation's coastal zone". The development of the land can strongly affect the sea, for example the engineering of structures versus non-structures and the effects of erosion along the shore.

INTEGRATED COASTAL ZONE MANAGEMENT

Integrated coastal zone management means the integration of all aspects of the coastal zone; this includes environmentally, socially, culturally politically and economically to meet a sustainable balance all around. Sustainability is the goal to allow development yet protect the environment in which we develop. Coastal zones are fragile and do not do well with change so it is important to acquire sustainable development. The integration from all views will entitle a holistic view for the best implementation and management of that country, region and local scales. The five types of integration include integration among sectors, integration between land and water elements of the coastal zone, integration amount levels of government, integration between nations and integration among disciplines are all essential to meet the needs for implementation. Management practices include

1. maintaining the functional integrity of the coastal resource systems,without disrupting the environment
2. reducing resource-use conflicts, by making sure resources are used adequately and sustainably,
3. maintaining the health of the environment, which means to protect the ecosystems and natural cycle,
4. facilitating the progress of multisectoral development, which means allowing developers to develop within standards.

These four management practices should be based on a bottom-up approach, meaning the approach starts from a local level which is more intimate to the specific environment of that area. After assessment from the local level, the state and federal input can be implemented. The bottom-up approach is key for protecting the local environments because there is a diversity of environments that have specific needs all over the world.

ADAPTIVE MANAGEMENT

Adaptive management is another practice of development adaptation with the environment. Resources are the major factor when managing adaptively to a certain environment to accommodate all the needs of development and ecosystems. Strategies used must be flexible by either passive or active adaptive management include these key features:

- AIterative decision-making (evaluating results and adjusting actions on the basis of what has been learned)
- Feedback between monitoring and decisions (learning process)

- Explicit characterization of system uncertainty through multi-model inference (experimentation)
- Embracing risk and uncertainty as a way of building understanding (trial and error)

To achieve adaptive management is testing the assumptions to achieve a desired outcome, such as trial and error, find the best known strategy then monitoring it to adapt to the environment, and learning the outcomes of success and failures of a project.

MITIGATION

The purpose of mitigation is not only to minimize the loss of property damage, but minimize environmental damages due to development. To avoid impacts by not taking or limiting actions, to reduce or rectify impacts by rehabilitation or restoring the affected environments or instituting long-term maintenance operations and compensating for impacts by replacing or providing substitute environments for resources Structural mitigation is the current solution to eroding beaches and movement of sand is the use of engineered structures along the coast have been short lived and are only an illusion of safety to the public that result in long term damage of the coastline. Structural management deals with the use of the following: groins which are man-made solution to longshore current movements up and down the coast.

The use of groins are efficient to some extent yet cause erosion and sand build up father down the beaches. Bulkheads are man-made structures that help protect the homes built along the coast and other bodies of water that actually induce erosion in the long run.

Jetties are structures built to protect sand movement into the inlets where boats for fishing and recreation move through. The use of nonstructural mitigation is the practice of using organic and soft structures for solutions to protect against coastal hazards. These include: artificial dunes, which are used to create dunes that have been either developed on or eroded. There needs to be at least two lines of dunes before any development can occur. Beach Nourishment is a major source of nonstructural mitigation to ensure that beaches are present for the communities and for the protection of the coastline. Vegetation is a key factor when protecting from erosion, specifically for to help stabilize dune erosion.

MARINE POLLUTION

Marine pollution occurs when harmful, or potentially harmful, effects result from the entry into the ocean of chemicals, particles, industrial, agricultural and residential waste, noise, or the spread of invasive organisms. Most sources of marine pollution are land based. The pollution often comes from nonpoint sources such as agricultural runoff and wind blown debris and dust. Nutrient pollution, a form of water pollution, refers to contamination by excessive inputs of nutrients. It is a primary cause of eutrophication of surface waters, in which excess nutrients, usually nitrogen or phosphorus, stimulate algal growth. Many potentially toxic chemicals adhere to tiny particles which are then taken up by plankton and benthos animals, most of which are either deposit or filter feeders. In this way, the toxins are concentrated upward within ocean food chains. Many particles combine chemically in a manner highly depletive of oxygen, causing estuaries to become anoxic.

When pesticides are incorporated into the marine ecosystem, they quickly become absorbed into marine food webs. Once in the food webs, these pesticides can cause mutations, as well as diseases, which can be harmful to humans as well as the entire food web. Toxic metals can also be introduced into marine food webs. These can cause a change to tissue matter, biochemistry, behaviour, reproduction, and suppress growth in marine life. Also, many animal feeds have a high fish meal or fish hydrolysate content. In this way, marine toxins can be transferred to land animals, and appear later in meat and dairy products.

HISTORY

Although marine pollution has a long history, significant international laws to counter it were only enacted in the twentieth century. Marine pollution was a concern during several United Nations Conferences on the Law of the Sea beginning in the 1950s. Most scientists believed that the oceans were so vast that they had unlimited ability to dilute, and thus render pollution, harmless. In the late 1950s and early 1960s, there were several controversies about dumping radioactive waste off the coasts of the United States by companies licensed by the Atomic Energy Commission, into the Irish Sea from the British reprocessing facility at Windscale, and into the Mediterranean Sea by the French Commissariat à l'Energie Atomique. After the Mediterranean Sea controversy, for example, Jacques Cousteau became a worldwide figure in the campaign to stop marine pollution. Marine pollution made further international headlines after the 1967 crash of the oil tanker Torrey Canyon, and after the 1969 Santa Barbara oil spill off the coast of California.

Marine pollution was a major area of discussion during the 1972 United Nations Conference on the Human Environment, held in Stockholm. That year also saw the signing of the Convention on the Prevention of Marine Pollution by Dumping of Wastes and Other Matter, sometimes called the London Convention. The London Convention did not ban marine pollution, but it established black and gray lists for substances to be banned (black) or regulated by national authorities (gray). Cyanide and high-level radioactive waste, for example, were put on the black list. The London Convention applied only to waste dumped from ships, and thus did nothing to regulate waste discharged as liquids from pipelines.

PATHWAYS OF POLLUTION

There are many different ways to categorize, and examine the inputs of pollution into our marine ecosystems. Patin (n.d.) notes that generally there are three main types of inputs of pollution into the ocean: direct discharge of waste into the oceans, runoff into the waters due to rain, and pollutants that are released from the atmosphere.

One common path of entry by contaminants to the sea are rivers. The evaporation of water from oceans exceeds precipitation. The balance is restored by rain over the continents entering rivers and then being returned to the sea. The Hudson in New York State and the Raritan in New Jersey, which empty at the northern and southern ends of Staten Island, are a source of mercury contamination of zooplankton (copepods) in the open ocean. The highest concentration in the filter-feeding copepods is not at the mouths of these rivers but 70 miles south, nearer Atlantic City, because water flows close to the coast. It takes a few days before toxins are taken up by the plankton.

Pollution is often classed as point source or nonpoint source pollution. Point source pollution occurs when there is a single, identifiable, and localized source of the pollution. An example is directly discharging sewage and industrial waste into the ocean. Pollution such as this occurs particularly in developing nations. Nonpoint source pollution occurs when the pollution comes from ill-defined and diffuse sources. These can be difficult to regulate. Agricultural runoff and wind blown debris are prime examples.

DIRECT DISCHARGE

Pollutants enter rivers and the sea directly from urban sewerage and industrial waste discharges, sometimes in the form of hazardous and toxic wastes. Inland mining for copper, gold. etc., is another source of marine pollution. Most of the pollution is simply soil, which ends up in rivers flowing to the sea. However, some minerals discharged in the course of the mining can cause problems, such as copper, a common industrial pollutant, which can interfere with the life history and development of coral polyps. Mining has a poor environmental track record. For example, according to the United States Environmental Protection Agency, mining has contaminated portions of the headwaters of over 40% of watersheds in the western continental US. Much of this pollution finishes up in the sea.

LAND RUNOFF

Surface runoff from farming, as well as urban runoff and runoff from the construction of roads, buildings, ports, channels, and harbours, can carry soil and particles laden with carbon, nitrogen, phosphorus, and minerals. This nutrient-rich water can cause fleshy algae and phytoplankton to thrive in coastal areas; known as algal blooms, which have the potential to create hypoxic conditions by using all available oxygen.

Polluted runoff from roads and highways can be a significant source of water pollution in coastal areas. About 75 percent of the toxic chemicals that flow into Puget Sound are carried by stormwater that runs off paved roads and driveways, rooftops, yards and other developed land.

SHIP POLLUTION

Ships can pollute waterways and oceans in many ways. Oil spills can have devastating effects. While being toxic to marine life, polycyclic aromatic hydrocarbons (PAHs), found in crude oil, are very difficult to clean up, and last for years in the sediment and marine environment.

Discharge of cargo residues from bulk carriers can pollute ports, waterways and oceans. In many instances vessels intentionally discharge illegal wastes despite foreign and domestic regulation prohibiting such actions. It has been estimated that container ships lose over 10,000 containers at sea each year (usually during storms). Ships also create noise pollution that disturbs natural wildlife, and water from ballast tanks can spread harmful algae and other invasive species.

Ballast water taken up at sea and released in port is a major source of unwanted exotic marine life. The invasive freshwater zebra mussels, native to the Black, Caspian and Azov seas, were probably transported to the Great Lakes via ballast water from a transoceanic vessel. Meinesz believes that one of the worst cases of a single invasive species causing harm to an ecosystem can be attributed to a seemingly harmless jellyfish. *Mnemiopsis leidyi*, a species of comb jellyfish that spread so it now inhabits estuaries in many parts of the world. It was first introduced in 1982, and thought to have been transported to the Black Sea in a ship's ballast water. The population of the jellyfish shot up exponentially and, by 1988, it was wreaking havoc upon the local fishing industry. "The anchovy catch fell from 204,000 tons in 1984 to 200 tons in 1993; sprat from 24,600 tons in 1984 to 12,000 tons in 1993; horse mackerel from 4,000 tons in 1984 to zero in 1993." Now that the jellyfish have exhausted the zooplankton, including fish larvae, their numbers have fallen dramatically, yet they continue to maintain a stranglehold on the ecosystem.

Invasive species can take over once occupied areas, facilitate the spread of new diseases, introduce new genetic material, alter underwater seascapes and jeopardize the ability of native species to obtain food. Invasive species are responsible for about $138 billion annually in lost revenue and management costs in the US alone.

ATMOSPHERIC POLLUTION

Another pathway of pollution occurs through the atmosphere. Wind blown dust and debris, including plastic bags, are blown seaward from landfills and other areas. Dust from the Sahara moving around the southern periphery of the subtropical ridge moves into the Caribbean and Florida during the warm season as the ridge builds and moves northward through the subtropical Atlantic. Dust can also be attributed to a global transport from the Gobi and Taklamakan deserts across Korea, Japan, and the Northern Pacific to the Hawaiian Islands. Since 1970, dust outbreaks have worsened due to periods of drought in Africa. There is a large variability in dust transport to the Caribbean and Florida from year to year; however, the flux is greater during positive phases of the North Atlantic Oscillation. The USGS links dust events to a decline in the health of coral reefs across the Caribbean and Florida, primarily since the 1970s.

Climate change is raising ocean temperatures and raising levels of carbon dioxide in the atmosphere. These rising levels of carbon dioxide are acidifying the oceans. This, in turn, is altering aquatic ecosystems and modifying fish distributions, with impacts on the sustainability of fisheries and the livelihoods of the communities that depend on them. Healthy ocean ecosystems are also important for the mitigation of climate change.

DEEP SEA MINING

Deep sea mining is a relatively new mineral retrieval process that takes place on the ocean floor. Ocean mining sites are usually around large areas of polymetallic nodules or active and extinct hydrothermal vents at about 1,400 - 3,700 meters below the ocean's surface. The vents create sulfide deposits, which contain precious metals such as silver, gold, copper, manganese, cobalt, and zinc. The deposits are mined using either hydraulic pumps or bucket systems that take ore to the surface to be processed. As with all mining operations, deep sea mining raises questions about environmental damages to the surrounding areas

Because deep sea mining is a relatively new field, the complete consequences of full scale mining operations are unknown. However, experts are certain that removal of parts of the sea floor will result in disturbances to the benthic layer, increased toxicity of the water column and sediment plumes from tailings. Removing parts of the sea floor disturbs the habitat of benthic organisms, possibly, depending on the type of mining and location, causing permanent disturbances. Aside from direct impact of mining the area, leakage, spills and corrosion would alter the mining area's chemical makeup. Among the impacts of deep sea mining, sediment plumes could have the greatest impact.

Plumes are caused when the tailings from mining (usually fine particles) are dumped back into the ocean, creating a cloud of particles floating in the water. Two types of plumes occur: near bottom plumes and surface plumes. Near bottom plumes occur when the tailings are pumped back down to the mining site. The floating particles increase the turbidity, or cloudiness, of the water, clogging filter-feeding apparatuses used by benthic organisms. Surface plumes cause a more serious problem. Depending on the size of the particles and water currents the plumes could spread over vast areas. The plumes could impact zooplankton and light penetration, in turn affecting the food web of the area.

TYPES OF POLLUTION

ACIDIFICATION

The oceans are normally a natural carbon sink, absorbing carbon dioxide from the atmosphere. Because the levels of atmospheric carbon dioxide are increasing, the oceans are becoming more acidic. The potential consequences of ocean acidification are not fully understood, but there are concerns that structures made of calcium carbonate may become vulnerable to dissolution, affecting corals and the ability of shellfish to form shells.

Oceans and coastal ecosystems play an important role in the global carbon cycle and have removed about 25% of the carbon dioxide emitted by human activities between 2000 and 2007 and about half the anthropogenic CO_2 released since the start of the industrial revolution. Rising ocean temperatures and ocean acidification means that the capacity of the ocean carbon sink will gradually get weaker, giving rise to global concerns expressed in the Monaco and Manado Declarations.

A report from NOAA scientists published in the journal Science in May 2008 found that large amounts of relatively acidified water are upwelling to within four miles of the Pacific continental shelf area of North America. This area is a critical zone where most local marine life lives or is born. While the paper dealt only with areas from Vancouver to northern California, other continental shelf areas may be experiencing similar effects. A related issue is the methane clathrate reservoirs found under sediments on the ocean floors. These trap large amounts of the greenhouse gas methane, which ocean warming has the potential to release. In 2004 the global inventory of ocean methane clathrates was estimated to occupy between one and five million cubic kilometres. If all these clathrates were to be spread uniformly across the ocean floor, this would translate to a thickness between three and fourteen metres. This estimate corresponds to 500-2500 gigatonnes carbon (Gt C), and can be compared with the 5000 Gt C estimated for all other fossil fuel reserves.

EUTROPHICATION

Eutrophication is an increase in chemical nutrients, typically compounds containing nitrogen or phosphorus, in an ecosystem. It can result in an increase in the ecosystem's primary productivity (excessive plant growth and decay), and further effects including lack of oxygen and severe reductions in water quality, fish, and other animal populations. The biggest culprit are rivers that empty into the ocean, and with it the many chemicals used as fertilizers in agriculture as well as waste from livestock and humans. An excess of oxygen depleting chemicals in the water can lead to hypoxia and the creation of a dead zone. Estuaries tend to be naturally eutrophic because land-derived nutrients are concentrated where runoff enters the marine environment in a confined channel. The World Resources Institute has identified 375 hypoxic coastal zones around the world, concentrated in coastal areas in Western Europe, the Eastern and Southern coasts of the US, and East Asia, particularly in Japan. In the ocean, there are frequent red tide algae blooms that kill fish and marine mammals and cause respiratory problems in humans and some domestic animals when the blooms reach close to shore.

In addition to land runoff, atmospheric anthropogenic fixed nitrogen can enter the open ocean. A study in 2008 found that this could account for around one third of the ocean's external (non-recycled) nitrogen supply and up to three per cent of the annual new marine biological production. It has been suggested that accumulating reactive nitrogen in the environment may have consequences as serious as putting carbon dioxide in the atmosphere. One proposed solution to eutrophication in estuaries is to restore shellfish populations, such as oysters. Oyster reefs remove nitrogen from the water column and filter out suspended solids, subsequently reducing the likelihood or extent of harmful algal blooms or anoxic conditions. Filter feeding activity is considered beneficial to water quality by controlling phytoplanton density and sequestering nutrients, which can be removed from the system through shellfish harvest, buried in the sediments, or lost through denitrification. Foundational work toward the idea of improving marine water quality through shellfish cultivation to was conducted by Odd Lindahl et al., using mussels in Sweden.

PLASTIC DEBRIS

Marine debris is mainly discarded human rubbish which floats on, or is suspended in the ocean. Eighty percent of marine debris is plastic - a component that has been rapidly accumulating since the end of World War II. The mass of plastic in the oceans may be as high as one hundred million metric tons. Discarded plastic bags, six pack rings and other forms of plastic waste which finish up in the ocean present dangers to wildlife and fisheries. Aquatic life can be threatened through entanglement, suffocation, and ingestion. Fishing nets, usually made of plastic, can be left or lost in the ocean by fishermen. Known as ghost nets, these entangle fish, dolphins, sea turtles, sharks, dugongs, crocodiles, seabirds, crabs, and other creatures, restricting movement, causing starvation, laceration and infection, and, in those that need to return to the surface to breathe, suffocation. Many animals that live on or in the sea consume flotsam by mistake, as it often looks similar to their natural prey. Plastic debris, when bulky or tangled, is difficult to pass, and may become permanently lodged in the digestive tracts of these animals, blocking the passage of food and causing death through starvation or infection. Plastics accumulate because they don't biodegrade in the way many other substances do. They will photodegrade on exposure to the sun, but they do so properly only under dry conditions, and water inhibits this process. In marine environments, photodegraded plastic disintegrates into ever smaller pieces while remaining polymers, even down to the molecular level. When floating plastic particles photodegrade down to zooplankton sizes, jellyfish attempt to consume them, and in this way the plastic enters the ocean food chain. Many of these long-lasting pieces end up in the stomachs of marine birds and animals, including sea turtles, and black-footed albatross.

Plastic debris tends to accumulate at the centre of ocean gyres. In particular, the Great Pacific Garbage Patch has a very high level of plastic particulate suspended in the upper water column. In samples taken in 1999, the mass of plastic exceeded that of zooplankton (the dominant animal life in the area) by a factor of six. Midway Atoll, in common with all the Hawaiian Islands, receives substantial amounts of debris from the garbage patch. Ninety percent plastic, this debris accumulates on the beaches of Midway where it becomes a hazard to the bird population of the island. Midway Atoll is home to two-thirds (1.5 million) of the global population of Laysan Albatross. Nearly all of these albatross have plastic in their digestive system and one-third of their chicks die.

Toxic additives used in the manufacture of plastic materials can leach out into their surroundings when exposed to water. Waterborne hydrophobic pollutants collect and magnify on the surface of plastic debris, thus making plastic far more deadly in the ocean than it would be on land. Hydrophobic contaminants are also known to bioaccumulate in fatty tissues, biomagnifying up the food chain and putting pressure on apex predators. Some plastic additives are known to disrupt the endocrine system when consumed, others can suppress the immune system or decrease reproductive rates. Floating debris can also absorb persistent organic pollutants from seawater, including PCBs, DDT and PAHs. Aside from toxic effects, when ingested some of these are mistaken by the animal brain for estradiol, causing hormone disruption in the affected wildlife.

TOXINS

Apart from plastics, there are particular problems with other toxins that do not disintegrate rapidly in the marine environment. Examples of persistent toxins are PCBs, DDT, pesticides, furans, dioxins, phenols and radioactive waste. Heavy metals are metallic chemical elements that have a relatively high density and are toxic or poisonous at low concentrations. Examples are mercury, lead, nickel, arsenic and cadmium. Such toxins can accumulate in the tissues of many species of aquatic life in a process called bioaccumulation. They are also known to accumulate in benthic environments, such as estuaries and bay muds: a geological record of human activities of the last century.

Specific examples

- Chinese and Russian industrial pollution such as phenols and heavy metals in the Amur River have devastated fish stocks and damaged its estuary soil.
- Wabamun Lake in Alberta, Canada, once the best whitefish lake in the area, now has unacceptable levels of heavy metals in its sediment and fish.
- Acute and chronic pollution events have been shown to impact southern California kelp forests, though the intensity of the impact seems to depend on both the nature of the contaminants and duration of exposure.
- Due to their high position in the food chain and the subsequent accumulation of heavy metals from their diet, mercury levels can be high in larger species such as bluefin and albacore. As a result, in March 2004 the United States FDA issued guidelines recommending that pregnant women, nursing mothers and children limit their intake of tuna and other types of predatory fish.
- Some shellfish and crabs can survive polluted environments, accumulating heavy metals or toxins in their tissues. For example, mitten crabs have a remarkable ability to survive in highly modified aquatic habitats, including polluted waters. The farming and harvesting of such species needs careful management if they are to be used as a food.
- Surface runoff of pesticides can alter the gender of fish species genetically, transforming male into female fish.
- Heavy metals enter the environment through oil spills - such as the Prestige oil spill on the Galician coast - or from other natural or anthropogenic sources.
- In 2005, the 'Ndrangheta, an Italian mafia syndicate, was accused of sinking at least 30 ships loaded with toxic waste, much of it radioactive. This has led to widespread investigations into radioactive-waste disposal rackets.
- Since the end of World War II, various nations, including the Soviet Union, the United Kingdom, the United States, and Germany, have disposed of chemical weapons in the Baltic Sea, raising concerns of environmental contamination.

UNDERWATER NOISE

Marine life can be susceptible to noise or sound pollution from sources such as passing ships, oil exploration seismic surveys, and naval low-frequency active sonar. Sound travels more rapidly and over larger distances in the sea than in the atmosphere. Marine animals, such as cetaceans, often have weak eyesight, and live in a world largely defined by acoustic information. This applies also to many deeper sea fish, who live in a world of darkness. Between 1950 and 1975, ambient noise in the ocean increased by about ten decibels (that is a tenfold increase).

Noise also makes species communicate louder, which is called the Lombard vocal response. Whale songs are longer when submarine-detectors are on. If creatures don't "speak" loud enough, their voice can be masked by anthropogenic sounds. These unheard voices might be warnings, finding of prey, or preparations of net-bubbling. When one species begins speaking louder, it will mask other species voices, causing the whole ecosystem to eventually speak louder.

According to the oceanographer Sylvia Earle, "Undersea noise pollution is like the death of a thousand cuts. Each sound in itself may not be a matter of critical concern, but taken all together, the noise from shipping, seismic surveys, and military activity is creating a totally different environment than existed even 50 years ago. That high level of noise is bound to have a hard, sweeping impact on life in the sea."

Index

M

O

P

R

S

T

U

V

W

Z